The SOUND *of a* WILD SNAIL EATING

Center Point
Large Print

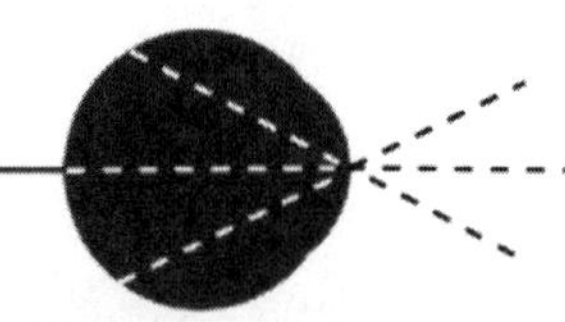

The SOUND of a WILD SNAIL EATING

Elisabeth Tova Bailey

CENTER POINT LARGE PRINT
THORNDIKE, MAINE

This Center Point Large Print edition is
published in the year 2011 by arrangement with
Algonquin Books of Chapel Hill,
a division of Workman Publishing.

This book is based on an essay that appeared in
The Missouri Review. Permissions for other material
reprinted in this book appear on pages 183–187,
which constitute an extension of the copyright page.

The text of this Large Print edition is unabridged.
In other aspects, this book may vary
from the original edition.
Printed in the United States of America.
Set in 16-point Times New Roman type.

ISBN: 978-1-61173-020-3

Library of Congress Cataloging-in-Publication Data

Bailey, Elisabeth Tova.
The sound of a wild snail eating / Elisabeth Tova Bailey. — Center
Point large print ed.
p. cm.
Originally published: Chapel Hill, N.C. : Algonquin Books of Chapel
Hill, 2010.
ISBN 978-1-61173-020-3 (lg. print : lib. bdg. : alk. paper)
1. Snails as pets—Anecdotes. 2. Gastropoda—Physiology.
3. Gastropoda—Anatomy. 4. Bailey, Elisabeth Tova—Health.
5. Chronically ill—Biography. I. Title.
SF459.S48B35 2011
594′.38—dc22

2010052179

To biophilia

A small pet is often an excellent companion.

—FLORENCE NIGHTINGALE, *Notes on Nursing,* 1912

The natural world is the refuge of the spirit . . . richer even than human imagination.

—EDWARD O. WILSON, *Biophilia,* 1984

Table of Contents

Prologue

Viruses are embedded into the very fabric of all life.
—Luis P. Villarreal,
"The Living and Dead Chemical Called a Virus," 2005

From my hotel window I look over the deep glacial lake to the foothills and the Alps beyond. Twilight vanishes the hills into the mountains; then all is lost to the dark.

After breakfast, I wander the cobbled village streets. The frost is out of the ground, and huge bushes of rosemary bask fragrantly in the sun. I take a trail that meanders up the steep, wild hills past flocks of sheep. High on an outcrop, I lunch on bread and cheese. Late in the afternoon along the shore, I find ancient pieces of pottery, their edges smoothed by waves and time. I hear that a virulent flu is sweeping this small town.

A few days pass and then comes a delirious

night. My dreams are disturbed by the comings and goings of ferries. Passengers call into the dark, startling me awake. Each time I fall back into sleep, the lake's watery sound pulls at me. Something is wrong with my body. Nothing feels right.

In the morning I am weak and can't think. Some of my muscles don't work. Time becomes strange. I get lost; the streets go in too many directions. The days drift past in confusion. I pack my suitcase, but for some reason it's impossible to lift. It seems to be stuck to the floor. Somehow I get to the airport. Seated next to me on the transatlantic flight is a sick surgeon; he sneezes and coughs continually. My rare, much-needed vacation has not gone as planned. I'll be okay; I just want to get home.

After a flight connection in Boston, I land at my small New England airport near midnight. In the parking lot, as I bend over to dig my car out of the snow, the shovel turns into a crutch that I use to push myself upright. I don't know how I get home. Arising the next morning, I immediately faint to the floor. Ten days of fever with a pounding headache. Emergency room visits. Lab tests. I am sicker than I have ever been. Childhood pneumonia, college mononucleosis—those were nothing compared to this.

A few weeks later, resting on the couch, I spiral into a deep darkness, falling farther and farther away until I am impossibly distant. I cannot come back up; I cannot reach my body. Distant sound of an ambulance siren. Distant sound of doctors talking. My eyelids heavy as boulders. I try to open them to a slit, just for a few seconds, but they close against my will. All I can do is breathe.

The doctors will know how to fix me. They will stop this. I keep breathing. What if my breath stops? I need to sleep, but I am afraid to sleep. I try to watch over myself; if I go to sleep, I might never wake up again.

Part 1

The Violet-pot Adventures

Try to love the questions themselves *as if they were locked rooms or books written in a very foreign language. Don't search for the answers, which could not be given to you now, because you would not be able to live them. And the point is, to live everything.* Live *the questions now.*

—Rainer Maria Rilke, 1903,
from *Letters to a Young Poet*, 1927

1. Field Violets

at my feet
when did you get here?
snail

— KOBAYASHI ISSA (1763–1828)

In early spring, a friend went for a walk in the woods and, glancing down at the path, saw a snail. Picking it up, she held it gingerly in the palm of her hand and carried it back toward the studio where I was convalescing. She noticed some field violets on the edge of the lawn. Finding a trowel, she dug a few up, then planted them in a terra-cotta pot and placed the snail beneath their leaves. She brought the pot into the studio and put it by my bedside.

"I found a snail in the woods. I brought it back and it's right here beneath the violets."

"You did? Why did you bring it in?"

"I don't know. I thought you might enjoy it."

"Is it alive?"

She picked up the brown acorn-sized shell and looked at it.

"I think it is."

Why, I wondered, would I *enjoy* a snail? What on earth would I do with it? I couldn't get out of bed to return it to the woods. It was not of much interest, and if it *was* alive, the responsibility—especially for a snail, something so uncalled for—was overwhelming.

My friend hugged me, said good-bye, and drove off.

At age thirty-four, on a brief trip to Europe, I was felled by a mysterious viral or bacterial pathogen, resulting in severe neurological symptoms. I had thought I was indestructible. But I wasn't. If anything did go wrong, I figured modern medicine would fix me. But it didn't. Medical specialists at several major clinics couldn't diagnose the infectious culprit. I was in and out of the hospital for months, and the complications were life threatening. An experimental drug that became available stabilized my condition, though it would be several grueling years to a partial recovery and a return to work. My doctors said the illness was behind me, and I wanted to believe them. I was ecstatic to have most of my life back.

But out of the blue came a series of insidious relapses, and once again, I was bedridden. Further, more sophisticated testing showed that the mitochondria in my cells no longer functioned correctly and there was damage to my autonomic nervous system; all functions not consciously directed, including heart rate, blood pressure, and digestion, had gone haywire. The drug that had previously helped now caused dangerous side effects; it would soon be removed from the market.

When the body is rendered useless, the mind still runs like a bloodhound along well-worn trails of neurons, tracking the echoing questions: the confused family of *why*s, *what*s, and *when*s and their impossibly distant kin *how.* The search is exhaustive; the answers, elusive. Sometimes my mind went blank and listless; at other times it was flooded with storms of thought, unspeakable sadness, and intolerable loss.

Given the ease with which health infuses life with meaning and purpose, it is shocking how swiftly illness steals away those certainties. It was all I could do to get through each moment, and each moment felt like an endless hour, yet days slipped silently past. Time unused and only endured still vanishes, as if time itself is starving, and each day is swallowed whole, leaving no crumbs, no memory, no trace at all.

I had been moved to a studio apartment where I could receive the care I needed. My own farmhouse, some fifty miles away, was closed up. I did not know if or when I'd ever make it home again. For now, my only way back was to close my eyes and remember. I could see the early spring there, the purple field violets—like those at my bedside—running rampant through the yard. And the fragrant small pink violets that I had planted in the little woodland garden to the north of my house—they, too, would be in bloom. Though not usually hardy this far north, somehow they survived. In my mind I could smell their sweetness.

Before my illness, my dog, Brandy, and I had often wandered the acres of forest that stretched beyond the house to a hidden, mountain-fed brook. The brook's song of weather and season followed us as we crisscrossed its channel over partially submerged boulders. On the trail home, in the boggiest of spots, perched on tiny islands of root and moss, I found diminutive wild white violets, their throats faintly striped with purple.

These field violets in the pot at my bedside were fresh and full of life, unlike the usual cut flowers brought by other friends. Those lasted just a few days, leaving murky, odoriferous vase water. In my twenties I had earned my living as a gardener, so I was glad to have this bit of garden right by my bed. I could even water the violets with my drinking glass.

But what about this snail? What would I do with it? As tiny as it was, it had been going about its day when it was picked up. What right did my friend and I have to disrupt its life? Though I couldn't imagine what kind of life a snail might lead.

I didn't remember ever having noticed any snails on my countless hikes in the woods. Perhaps, I thought, looking at the nondescript brown creature, it was precisely because they were so inconspicuous. For the rest of the day the snail stayed inside its shell, and I was too worn out from my friend's visit to give it another thought.

2. Discovery

the snail gets up
and goes to bed
with very little fuss
—KOBAYASHI ISSA (1763–1828)

Around dinnertime I was surprised to see that the snail was partway out of its shell. It was alive. The visible part of its body was nearly two inches long from head to tail, and moist. The rest of it was hidden in the attached inch-high brown shell, which it balanced gracefully on its back. I watched as it moved slowly down the side of the flowerpot. As it glided along, it gently waved the tentacles on its head.

Throughout the evening the snail explored the sides of the pot and the dish beneath. Its leisurely pace was mesmerizing. I wondered if it would wander off during the night. Perhaps I'd never see it again, and the snail problem would simply vanish.

But when I woke the next morning, the snail was back up in the pot, tucked into its shell, asleep beneath a violet leaf. The night before, I had propped an envelope containing a letter against the base of the lamp. Now I noticed a mysterious square hole just below the return address. This was baffling. How could a hole—a *square* hole—appear in an envelope overnight? Then I thought of the snail and its evening activity. The snail was clearly nocturnal. It must have some kind of teeth, and it wasn't shy about using them.

My healthy life had been full of activity, filled with friends, family, and work; the pleasures of gardening, hiking, and sailing; and the familiar humdrum of daily routines: making breakfast, exploring the woods, going to work, reading a book, getting up to get something. Now, getting up to get something, anything—that alone would be an accomplishment. From where I lay, all of life was out of reach.

As the months drifted by, it was hard to remember why the endless details of a healthy life and a good job had seemed so critical. It was odd to see my friends overwhelmed by their busy lives, when they could do all the things I could not, without a second thought.

Whereas the future had once beckoned with many intriguing paths, now there was just one impossible route. So it was into the past, with its rich sedimentary layers, that my mind would go instead. A breath of wind through an open window stirred the memory of crossing Penobscot Bay on the bowsprit of a schooner. With the simple wish to brush my teeth came thoughts of my farmhouse bathroom, with its window view of the old apple trees and the poppy garden. It had amused me to see the laundry hanging on its line over the poppies; their yellows, oranges, and reds accented the blue sheets and the night-gowns, which reached with their arms down toward the flowers.

On the second morning of the snail's stay, I found another square hole, this time in a list I was keeping on a scrap of paper. As each successive morning arrived, so did more holes. Their square shape continued to perplex me. Friends were surprised and amused to receive postcards with an arrow pointing at a hole and my scrawled note: "Eaten by my snail."

It dawned on me that perhaps the snail needed some real food. Letters and envelopes were probably not its typical diet. A few long-gone flowers were in a vase by my bed. One evening I put some of the withered blossoms in the dish beneath the pot of violets. The snail was awake. It made its way down the side of the pot and investigated the offering with great interest and then began to eat one of the blossoms. A petal started to disappear at a barely discernible rate. I listened carefully. I could *hear* it eating. The sound was of someone very small munching celery continuously. I watched, transfixed, as over the course of an hour the snail meticulously ate an entire purple petal for dinner.

The tiny, intimate sound of the snail's eating gave me a distinct feeling of companionship and shared space. It also pleased me that I could recycle the withered flowers by my bed to sustain a small creature in need. I might prefer my salad fresh, but the snail preferred its salad half-dead, for not once had it nibbled on the live violet plants that provided its sleeping shelter. One has to respect the preferences of another creature, no matter its size, and I did so gladly.

The studio apartment where I was staying had lots of windows and a beautiful view of a salt marsh. But the windows were far from where I lay, and I could not sit up to see out. Though they brought me light each day, the world they framed was beyond my reach. Unlike my own farmhouse, which was full of color, the walls and ceiling of this room where I woke each morning were entirely white—I felt trapped inside a stark white box.

During the earlier years of my illness, I had spent countless hours on a daybed in my 1830s farmhouse, staring up at the hand-hewn beams overhead. Their rich, golden brown hues soothed my soul; the knots told a history of branches and long-ago wilderness; the square-headed nails sticking out here and there once had purpose. Each room in the house was trimmed in an old-fashioned milk-paint color. In the room where I lay, the trim was a deep blue, and I could turn my head to see red in the kitchen, green in the bathroom, and a calm gray in the front room.

The daybed at home was right next to a window so that I could look out without sitting up. In the summer my perennial gardens were in view, untended but still thriving. I would watch for the arrival of friends as they came by foot, bike, or car, bringing stories to tell, and I'd wave them off as they set out again. When I woke each morning at dawn, several cats would be prowling the field. I'd hear my neighbors drive off to work, one by one. The slant of sun would slowly steepen toward midday, then lengthen as it slowly fell away. One by one my neighbors returned. Evening settled over the field, the cats took up their hunting in the long grass, and finally night descended.

Though I was grateful for the care I was receiving here in this white room, I was not at home. It was hard enough that my body was a bizarre and bewildering place, but I was homesick as well. I was far from the things that delighted me, the wild woods that sustained me, and the social network that enriched me.

Survival often depends on a specific focus: a relationship, a belief, or a hope balanced on the edge of possibility. Or something more ephemeral: the way the sun passes through the hard, seemingly impenetrable glass of a window and warms the blanket, or how the wind, invisible but for its wake, is so loud one can hear it through the insulated walls of a house.

For several weeks the snail lived in the flower-pot just inches from my bed, sleeping beneath the violet leaves by day and exploring by night. Each morning while I was having breakfast it climbed back into the pot to sleep in the little hollow it had made in the dirt. Though the snail usually slept through the days, it was comforting to glance toward the violets and see its small circular shape tucked under a leaf.

Each evening the snail awoke, and with astonishing poise, it moved gracefully to the rim of the pot and peered over, surveying, once again, the strange country that lay ahead. Pondering its circumstance with a regal air, as if from the turret of a castle, it waved its tentacles first this way and then that, as though responding to a distant melody.

As I prepared for the night, the snail moved in its leisurely way down the side of the pot to the dish beneath. It found the flower blossoms I had placed there and began its breakfast.

3. Explorations

As the exploration is pressed,
it will engage more of the things close
to the human heart and spirit.

—EDWARD O. WILSON, *Biophilia,* 1984

When I woke during the night, I would listen intently. Sometimes the silence was complete, but at other times I could hear the comforting sound of the snail's minuscule munching. With my flashlight I'd search until the beam of light found its small shape. If it was eating, I'd peek to see which of the wizened flowers it preferred. It usually stayed within a few feet of the flowerpot, which sat on a crate that I was using as a bedside table.

Every few days I watered the violets from my drinking glass, and the excess water seeped into the dish beneath. This always woke the snail. It would glide to the rim of the pot and look over, slowly waving its tentacles in apparent delight, before making its way down to the dish for a drink. Sometimes it started back up, only to stop at a halfway point and go to sleep. Waking periodically, and without moving from its position, it would stretch its neck all the way down to the water and take a long drink.

A little more dirt was needed around the roots of the violets, which my caregiver procured from the vegetable garden and added to the flowerpot. The snail was *not* pleased. For the next few days it carefully crept up the side of the pot and directly onto a violet leaf, never touching the garden soil, settling in for the day's snooze perched high in the crown of the plant. Rather abashed, I asked for more help, and the sandy garden soil was exchanged for humus from the snail's own woods. Soon the snail was sleeping beneath the violet leaves again in a soft new hollow.

In the 1920s, the crate beneath the pot of violets had traveled to Burma and back with the belongings of my maternal grandparents. They were medical missionaries, and my grandfather's skill as a doctor was well respected. He treated many people with illnesses and injuries and even saved the life of a man severely mauled by a tiger. When the sawbwa of Kengtung's favorite elephant was ailing, my grandfather was called. Bravely, he lanced the elephant's giant boil and treated the virulent infection.

My grandparents returned to New England, and my grandfather settled into life as a country doctor. The living room served as his office, and it was there that he saw patients. When I visited as a child, I was petrified he might hear me cough. A ticklish throat or the slightest pallor, and he'd rush to a large jar of revoltingly long tongue depressors, thrusting one down my gagging throat. Yet when he answered a patient's call, even in the middle of the night, his very first words were always "I am so sorry you are not feeling well." How rare it is to hear a doctor express such empathy.

As the weeks passed, the snail's nighttime forays became more adventurous, and so did its appetite. The flowers I fed it clearly were not enough. One night it ate part of the label on a vitamin C bottle. Another night it climbed up a pastel drawing made by an artist friend and ate some of the green border. I woke one morning to find a hole in a padded envelope for mailing books.

More and more frequently, in the middle of the night, the snail set off on a longer journey into new territory. I'd discover it partway down the side of the crate, sometimes nearly to the floor. Often, it investigated the india ink words stamped into the wood. It seemed to have a particular interest in anything the color of rich, dark soil, like the crate's black lettering or the base of the lamp. It was equally attracted to white things such as paper. Perhaps, I thought, paper was its woody version of fast food.

After being transported from the woods, the snail had emerged from its shell into the alien territory of my room, with no clue as to where it was or how it had arrived; the lack of vegetation and the desertlike surroundings must have seemed strange. The snail and I were both living in altered landscapes not of our choosing; I figured we shared a sense of loss and displacement.

Each morning there was a moment, before I had fully awakened, when my mind still groped its clumsy way back to consciousness, my body not yet remembered, reality not yet acknowledged. That moment was always full of pure, sweet, uncontrollable hope. I did not ask for this hope to come; I did not even want it, for it trailed disappointment in its wake. Yet there it was, hovering within me—hope that my illness had vanished with the night and my health had returned magically with daybreak. But that moment always passed, my eyes opened, and reality flooded in; nothing had changed at all.

Then I thought of the snail. I'd look for the tiny, earth-colored creature. Usually it was back up in the flowerpot asleep, its familiar shape reminding me that I wasn't alone.

By day, the strangeness of my situation was sharpest: I was bed-bound at a time when my friends and peers were moving forward in their careers and raising families. Yet the snail's daytime sleeping habits gave me a fresh perspective; I was not the only one resting away the days. The snail naturally slept by day, even on the sunniest of afternoons. Its companionship was a comfort to me and buffered my feelings of uselessness.

In the evenings there was a short but satisfying time when I knew the rest of the human world would join me, if just for the night, in my recumbent lifestyle. When healthy people take to their beds, they sink deeply into a privileged sleep. But with my illness, sleep was diaphanous and often nonexistent. The snail, once again, came to my rescue. As the world fell into sleep without me, the snail awoke, as if this darkest of times were indeed the *best* of times in which to live.

After weeks of around-the-clock companionship, there was no doubt about the relationship: the snail and I were officially cohabiting. I was, I admit, attached. I felt some guilt that it had been taken, unasked, from its natural habitat, yet I was not ready to part with it. It was adding a welcome focus to my life, and I couldn't think how I would otherwise have passed the hours.

Part 2

A GREEN KINGDOM

Think not of the amount to be accomplished, the difficulties to be overcome, or the end to be attained, but set earnestly at the little task at your elbow, letting that be sufficient for the day.

—SIR WILLIAM OSLER, physician (1849–1919)

4. The Forest Floor

I have set myself a goal, a certain rock,
but it may well be dawn
before I get there . . .
If and when I reach the rock,
I shall go into a certain crack
there for the night.

—ELIZABETH BISHOP,
from "Giant Snail," 1969

Despite its small size, the snail was a fearless and tireless explorer. Maybe it was searching for a trail back to its original woods or hoping to find better fare. Instinctively it knew its limits, how far it could travel during the night and still return home in the morning. On the crate's dry surface, the pot of violets was an oasis, offering water, food, and shelter.

Setting off on an expedition, its tentacles stretched out in anticipation, the snail appeared confident about where it was going, as if what it was looking for was just a few inches farther ahead. Watching it glide along was a welcome distraction and provided a sort of meditation; my often frantic and frustrated thoughts would gradually settle down to match its calm, smooth pace. With its mysterious, fluid movement, the snail was the quintessential tai chi master.

I began to worry about how far the snail might go in the night, the difficulties it might encounter in its travels, and what risky thing it might choose to sample for a meal. Ink, pastels, and label glues didn't seem like good forage for a snail. This brought to mind a children's verse from the A. A. Milne poem "The Four Friends," about an elephant, a lion, a goat, and a small snail named James. "James gave the huffle of a snail in danger / And nobody heard him at all." I didn't think that a snail could make the sound of a huffle—but I didn't want to find out.

Though the bed-and-breakfast arrangement in the flowerpot had worked for a while, I wanted the snail to have a safer and more natural home. There was a barn attached to the studio where I was staying, and in one of its dark corners my caregiver found an empty rectangular glass aquarium. This was soon converted into a roomy terrarium filled with fresh native plants and other materials from the snail's own woods: goldthread—aptly named for its colorful roots—holding its trio of delicate, paw-shaped leaves high on a thin stem; partridgeberry, with its round, dark green leaves and its small, bright red berries, which lasted for months; the larger, waxy leaves of checkerberry; many kinds of moss; small polypody ferns; a tiny spruce tree; a rotting birch log; and a piece of old bark encrusted with multicolored lichen.

Gulls flying over the coastline sometimes drop mussels, and in the woods one often finds the empty blue shells where they've landed in the moss. Such a shell, with its silvery inside, now served as a natural basin for fresh drinking water. With an old leaf here and a pine needle there, the terrarium looked as though a bit of native forest floor, with all its natural disarray, had been lifted up and placed inside. The moist, lush vibrancy of the plants reminded me of the woods after a rainstorm. It was a world fit for a snail, and it was a welcome sight for my own eyes as well.

Within moments of moving into this rich kingdom, the snail came partway out of its shell. Its tentacles quivered with interest and it set off to investigate the new terrain. It crawled along the dead log, drank water out of the mussel shell, investigated the mosses, climbed up the terrarium's glass side, and then chose a dark, private corner and went to sleep nestled in some moss.

While the snail slept, I explored the terrarium

from my bed, letting my eyes wander through the miniature hills and dales of its fresh green landscape. The variety of mosses was so satisfying, from a deep, loose softness to dense mounds with fuzzy and velvety textures. Their hues ranged from bright grass greens to deep dark greens and from sharp lemon greens to light blue greens.

Polypody ferns gently arched their beautiful four-inch fronds, their youngest fiddleheads still tightly curled. In my woods at home, along the brook, these ferns live on the sheer sides of granite boulders. They survive on a margin of rock where the air is humid and alive with the brook's energy, their rhizomes finding sustenance in cracks and crevices. Buried beneath winter's ice and snow each year, they magically send up new fronds every spring—a primeval perseverance.

The fresh terrarium at my side was lovely all by itself—a green and growing ecosystem; that it provided a magnificent backdrop for the humble brown snail was all the better. While the snail must have missed its familiar woods, the terrarium at least offered a more comfortable and natural world than the flowerpot. The snail would be safe in the terrarium, safer even than in the wild, as there were no predators hiding behind a leaf or swooping down from the sky.

As my snail watching continued, I wanted to

know more about how to care properly for my small companion. My caregiver unearthed a decades-old paperback book titled *Odd Pets*, by Dorothy Hogner. In addition to providing basic information on snails, Hogner suggested feeding them a diet of mushrooms.

There were some fresh portobellos in the kitchen refrigerator. A single portobello was about fifty times larger than my snail, and so my caregiver cut a generous slice and placed it in the terrarium. The snail loved the mushroom. It was so happy to have a familiar food, after weeks of nothing but wilted flowers, that for several days it slept right next to the huge piece of portobello, waking throughout the day to reach up and nibble before sinking back into a well-fed slumber. Each night a surprisingly large portion of the mushroom would vanish, until, by the end of the week, the very last piece had disappeared.

5. Life in a Microcosm

Everything in the world of Things
and animals
is still filled with happening,
which you can take part in.
—RAINER MARIA RILKE, 1903,
from *Letters to a Young Poet,* 1927

The snail consumed an entire slice of portobello every week. As I watched it eat, I noticed that it nodded its head gently up and down. Did this mean that it approved of its dinner? When I examined what remained of the mushroom after it had dined, I could see a pattern of fresh teeth marks—very fine little vertical striations, as if made by a tiny comb.

Half the fun of having the snail as a companion was that it kept finding new sleeping places. So there was an ongoing game of hide-and-seek in the terrarium. It would blend so well into the woodland plants that I'd have to sleuth out its latest hiding spot. If the day was cloudy or rainy, the snail awoke and was active, and I was amazed at how fast it moved. I'd see it in one place at one moment, and then my mind would wander off and I'd have to search the terrarium to find it again.

The creature seemed to defy physics. It moved over the very tips of mosses without bending them, and it could travel straight up the stem of a fern and then continue upside down along the frond's underside. Its tiny weight caused the fern frond to bend into an arc, yet the snail was unfazed; it was perfectly comfortable in any position and at any angle or height. Its balance, too, was impeccable. It could perch on the very edge of the mussel shell and from this precarious position reach casually across open space to eat some of the mushroom without falling or spilling water from the shell. No challenge was too great; if the snail came to an obstacle such as a branch, it made a brief inspection and then simply climbed up and over, rather than taking a longer route around. Each morning the terrarium glistened with the silvery trails of its nighttime travels.

I was fond of the elegant way the snail waved its tentacles as it moved serenely along, and I loved to watch it drink water from the mussel shell. Several times I was lucky enough to see it grooming; it arched its neck over the curved edge of its own shell and cleaned the rim carefully with its mouth, like a cat licking fur on the back of its neck. Usually the snail slept on its side, and at these times the striae, perpendicular to the spiraling whorls of its shell, reminded me of the pattern of stripes on my old tiger cat, Zephyr, when he would curl into a nap.

Though holding and reading a book for any length of time involved levels of strength and concentration that were beyond me, watching the snail was completely relaxing. I observed without thinking, looking into the terrarium simply to feel connected to another creature; another life was being lived just a few inches away.

While the snail and I each had our routines, we also both appreciated adventures. When a visiting friend or relative brought something to add to the terrarium, the snail was always intrigued. Whether it was a half-rotten lichen-covered branch, a piece of birch bark, a clump of moss of a different species, or perhaps a leaf of lettuce or a slice of cucumber, the snail received the gift with tentacles aquiver. After conducting a careful and thorough examination, it then tasted anything that might be edible.

My own adventures were more challenging. After weeks of never leaving the bed in the room where I stayed, a trip to a doctor's appointment was a monumental undertaking. I traveled horizontally in the car, and given the physical stillness of my usual daily existence, it was astonishing to see the treetops rushing past overhead at what seemed like a furious speed.

Wheeled into the doctor's reception room, I'd find myself surrounded by quietly waiting patients. We had each journeyed to this office from our own distant planet of illness. Though strangers, we became instant, silent companions. We were here for the same purpose: to describe our alien experience to the doctor in hope of survival advice. The chance to be with other patients brought a catch to my throat; despite our individual ailments, we shared the burden of illness. Yet even here my participation was limited, as I was too weak to sit upright for more than a few minutes. As quickly as possible I'd be taken straight back to an examination room so that I could wait lying down.

Though I could recline in the back of a car for these occasional outings, there were few other accessible destinations. Offices, stores, galleries, libraries, and movie theaters are not designed for horizontal people. The most satisfying adventure was when my driver had errands to run and I could lie in the back of the car in a parking lot and watch my own species bustle about its business. This brought a sense of connection and contentment, yet was a striking reminder of how entirely cut off I was from the most basic activities of life.

6. Time and Territory

The velocity of the ill, however,
is like that of the snail.

—EMILY DICKINSON,
in a letter to Charles H. Clark, April 1886

Inches from my bed and from each other stood the terrarium and a clock. While life in the terrarium flourished, time ticked away its seconds. But the relationship between time and the snail confused me. The snail would make its way through the terrarium while the hands of the clock hardly moved—so I often thought the snail traveled faster than time. Then, absorbed in snail watching, I'd find that time had flown by, unnoticed. And what about the unfurling of a fern frond? Its pace was undetectable, yet day by day it, too, reached toward its goal.

The mountain of things I felt I needed to do reached the moon, yet there was little I could do about anything, and time continued to drag me along its path. We are all hostages of time. We each have the same number of minutes and hours to live within a day, yet to me it didn't feel equally doled out. My illness brought me such an abundance of time that time was nearly all I had. My friends had so little time that I often wished I could give them what time I could not use. It was perplexing how in losing health I had gained something so coveted but to so little purpose.

I eagerly awaited visitors, but the anticipation and the extra energy of greetings caused a numbing exhaustion. As the first stories unfolded, my spirit held on to the conversation as best it could—I so wanted these connections to the outside world—but my body sank beneath waves of weakness. Still, my friends were golden threads randomly appearing in the monotonous fabric of my days. Each visit was a window that opened momentarily into the life I had once known, always falling shut before I could make my way back through. The visits were like dreams from which I awoke once more alone.

As the snail's world grew more familiar, my own human world became less so; my species was so large, so rushed, and so confusing. I found myself preoccupied with the energy level of my visitors, and I started to observe them in the same detail with which I observed the snail. The random way my friends moved around the room astonished me; it was as if they didn't know what to do with their energy. They were so *careless* with it. There were spontaneous gestures of their arms, the toss of a head, a sudden bend into a full body stretch as if it were nothing at all; or they might comb their fingers unnecessarily through their hair.

It took time for visitors to settle down. They sat and fidgeted for a while, then slowly relaxed until a calmness finally spread through them. They began to talk about more interesting things. But halfway through a visit, they would notice how little I moved, the stillness of my body, and an odd quietness would come over them. They would worry about wearing me out, but I could also see that I was a reminder of all they feared: chance, uncertainty, loss, and the sharp edge of mortality. Those of us with illnesses are the holders of the silent fears of those with good health.

Eventually, discomfort moved through my visitors, nudging a hand into motion, a foot into tapping. The more apparent my own lack of movement, the greater their need to move. Their energy would turn into restlessness, propelling their bodies into action with a flinging of the arms or a walk around the room; a body is not meant to be still. Soon my visitors were off.

My dog, Brandy, was a mix of golden retriever and yellow Lab. Even at eight years of age, her energy was extreme compared to my own. It was incredible that I, too, had once moved through life with such exuberance, with her at my side. From my bed I could give her scraps from my dinner and manage a few strokes of her soft ears. I loved her so, and her intense longing for more made me ache to leap from bed, fling open the door to the outside world, and escape, the two of us heading, once again, deep into the wild woods.

Whereas the energy of my human visitors wore me out, the snail inspired me. Its curiosity and grace pulled me further into its peaceful and solitary world. Watching it go about its life in the small ecosystem of the terrarium put me at ease. I began to think about naming the snail, as it was an individual with its own unique character traits. I had learned in the book *Odd Pets* that snails are hermaphrodites, which narrowed the options. But a human name didn't seem to fit. The snail was not just an individual creature that I was coming to know. It was introducing me in spirit to its entire line of gastropod ancestors, which, I guessed, reached far back in time. Looking into the terrarium was like entering that ancient era. From my recumbent bedside view, the ferns and mosses appeared as miniature forests and fields, and as I watched the snail go about its life, it seemed as if it lived in a timeless world without change. I liked the sound of the word "snail" every time I said it; the word was as small and simple as the creature itself. It is a word from Old English, with an earlier derivation from the German *schnecke*, for snail, spiral, or spiral-shaped yeast bun. So in the end I decided not to name my companion but to continue to refer to it as "the snail."

Given its tiny footprint, the snail had plenty of territory in the terrarium to survey in minute detail, finding endless nooks and crannies of interest. I, on the other hand, rarely moved beyond the familiar section of my sheets. Occasionally, when the snail slept and an urgent need for change—no matter the cost—swept through me, I would slowly roll from my right side over to my left side. This simple act caused my heart to beat wildly and erratically, but the reward was a whole new vista. The other side of the room was spread out before me like a map with countless possibilities of faraway adventures, including the most tantalizing of things, a window and a door.

Nothing, of course, was in reach. I could just see into the corner of the bathroom, where I knew, if I could only look farther in, I would find a claw-foot bathtub. Just to think of a bath, the kind one can settle into as if it were a relaxing, normal routine, caused an unfathomable longing. Across the room there was a shelf that held many books, each cover a different hue, their titles of possible interest if only I could decipher them, but the distance was too great. There was a window I could look out if only I could stand. And there was the door, the door to the outside world.

Was this truly a door that I would someday open and walk through, as if walking out into the world were an ordinary thing to do? I would look at the door until it reminded me of all the places I could not go. Then, exhausted and empty from my audacious adventure, I'd make the slow roll back toward the kingdom of the terrarium and the tiny life it contained.

Part 3

Juxtapositions

The history of the . . . snail has been more copiously considered than that of the elephant; and its anatomy is as well, if not better, known: however, not to give any one object more room in the general picture of nature than it is entitled to, it will be sufficient to observe that the snail is surprisingly fitted for the life it is formed to lead.

—Oliver Goldsmith,
A History of the Earth and Animated Nature, 1774

7. Thousands of Teeth

The mouth of the snail is armed
with a very formidable instrument
in the shape of a remarkable sword-like
tongue . . . [with an] immense number of
excessively sharp little teeth . . .
The quantity of these teeth is incredible.

—*Dietetic and Hygienic Gazette,* 1900

I thought my snail's tendency to gently nod its head as it ate was just a personality trait—but there was more to it than that. Years later I would read in more depth about the life of terrestrial snails. I requested through interlibrary loan the twelve-volume compendium The *Mollusca*, which covers the entire phylum of mollusks—soft-bodied creatures lacking a backbone—from the octopus with its humanlike intelligence down to the tiny snail.

The scientific name for a snail or slug—a

mollusk with a single muscular foot—is *gastropod;* derived from Latin and Greek, the word means "stomach-foot." The poet Billy Collins ends his wonderfully quirky poem "Evasive Maneuvers" with these lines:

> [I] said the word *gastropod* out loud,
> and having no idea what it meant
> went upstairs and looked it up
> then hid in the woods from my wife
>   and our dog.

If the term *gastropod* startled Collins, I wondered what surprises awaited me in the pages of *The Mollusca.* Arriving in random order, the dusty gray volumes were so heavy that I propped them up against other books and read them lying on my side. As I skimmed slowly along, reading a little bit each day, I found that every scientific field, from biology and physiology to ecology and paleontology, was packed with insights on gastropods. The abundance of detail was astonishing, ranging from their complex teeth patterns to the biochemistry of their slime making and the intimate details of their species-specific love lives. Yet even with *The Mollusca*'s many volumes, a certain perspective on snail life was missing. Then I discovered the nineteenth-century naturalists, intrepid souls who thought nothing of spending countless hours in the field observing their tiny subjects. I also came across poets and writers who had each, at some point in their life, become intrigued by the life of a snail.

In the fourth century BC, in the *History of Animals*, Aristotle noted that snail teeth are "sharp, and small, and delicate." My snail possessed around 2,640 teeth, so I'd add the word *plentiful* to Aristotle's description. The teeth point inward so as to give the snail a firm grasp on its food; with about 33 teeth per row and maybe eighty or so rows, they form a multi-toothed ribbon called a radula, which works much like a rasp. This explained my snail's nodding head as it grated away at a mushroom; it also explained the odd squareness of the holes I had discovered in my envelopes and lists. As the front row of teeth gets worn down, a fresh new row is added at the back and the radula slowly moves forward, being completely replaced over the course of four to six weeks. Radulae are adapted to a particular snail's diet and can be an identifying characteristic of a species.

With only thirty-two adult teeth, which had to last the rest of my life, I found myself experiencing tooth envy toward my gastropod companion. It seemed far more sensible to belong to a species that had evolved natural tooth replacement than to belong to one that had developed the dental profession. Nonetheless, dental appointments were one of my favorite adventures, as I could count on being recumbent. I could see myself settling into the dental chair, opening my mouth for my dentist, and surprising him with a human-sized radula.

Some snail species are predatory, and a few are even cannibalistic and will bore a hole through another snail's shell or attack directly through an aperture. These snails have evolved fewer but longer teeth, which, rather sinisterly, they can fold out of the way to give more mouth room for ingesting their victims.

This particular trait gave me the shudders. Even though my snail was not cannibalistic, I would not want to meet up with it or any other snail that was human-sized, which brought to mind Patricia Highsmith's short story "The Quest for *Blank Claveringi.*" Avery Clavering, a zoology professor, hears about the mythical man-eating snails of Kuwa, and hoping to prove their existence and gain fame by naming the species after himself, he sets off in pursuit. Arriving in Kuwa, he finds the giant twenty-foot snails grazing on treetops. Then they notice him. He assumes that it will be easy to "escape from two slow, lumbering creatures like the—the what? . . . *Carnivora* (perhaps) *Claveringi.*" But the professor gets a little too close to his live specimens, and so begins the most peculiar, and certainly the slowest, chase scene ever to occur in literature. With the giant snails in leisurely but relentless pursuit, Clavering becomes exhausted and seeks shelter between some boulders. One of the snails seals its slimy foot over his refuge, nearly smothering him. Eventually he escapes into the sea, but the giant snails follow and the plot reaches its dreadful end.

8. Telescopic Tentacles

The [snail's] tentacles are as expressive as a mule's ears, giving an appearance of listless enjoyment when they hang down, and an immense alertness if they are rigid, as happens when the snail is on a march.

—ERNEST INGERSOLL, "In a Snailery," 1881

When my snail was active, its muscular head and foot were extended, but at the slightest hint of a disturbance, it quickly withdrew them into the shell's largest, outermost whorl. Its soft body, containing the vital organs—a lung, a heart, and a gastrointestinal system—was connected to its shell by an internal mantle, which also provided space for a water reservoir. It could store about one-twelfth its weight in water and thus, camel-like, survive stretches of dry weather.

About half of my snail's respiration occurred through its skin, and the other half through a breathing pore—a little hole on its right side below its head. Called a pneumostome, this pore allows air exchange by diffusion, opening infrequently, maybe four times a minute, more or less, depending on the snail's activity. As warm-blooded creatures, or *homeotherms*, we humans have to maintain a constant body temperature, but my cold-blooded *poikilotherm* snail's temperature matched that of its changing environment. Thus it could get by on half the calories needed by a similarly sized mammal.

My snail was equipped with two pairs of tentacles: the lower pair were a quarter of an inch in length, while the upper pair were nearly half an inch long, with eyes at the tips. The snail could instantly retract its eyes through these hollow tentacles, which themselves retracted just as quickly into its head. "The first striking peculiarity [of the snail] is that the animal has got its eyes on the points of its largest horns," exclaimed Oliver Goldsmith in 1774 in *A History of the Earth and Animated Nature.* And at the end of the nineteenth century, in *The Dawn of Reason*, James Weir explained more precisely that "the snail carries its eyes in telescopic watchtowers."

When my snail was foraging for food or grazing on a mushroom, its tentacles quivered and twitched continuously. They stretched toward desirable smells but were instantly retracted from anything the least bit offensive. The snail could move its tentacles individually in nearly any direction up to a ninety-degree angle, sweeping them slowly back and forth and up and around, just as a boat under way in the dark swings its searchlights about to look for navigational marks.

While we humans have five senses, relying most heavily on vision to find our way, a snail relies almost entirely on just three senses: smell, taste, and touch, with smell being the most critical. My snail could not hear anything at all; it lived in a world of silence. Its "sight" was highly limited—just a general awareness of dark and light to help with orientation. Bright light might warn of a hotter, drier, and more challenging environment; dark suggested safer, cooler, more humid conditions. A sudden shadow might alert it to a predator.

It was its tentacles—which hold smell and taste receptors—that gave my snail its look of intelligence and purpose. So critical are they to a snail's survival that, if injured, they can be regrown, just as a starfish can regrow an arm. In an article titled "In the Realm of the Chemical," David H. Freedman explains:

> The land snail . . . devotes about half its . . . brain to taste and smell affairs. It divides the job neatly between its two pairs of

> [tentacles]: one [upper] pair is waved in the air to pick up smells, while the second [lower] pair is dipped tongue-style into promising substances as a final check before ingestion.

Using the taste buds on its lower tentacles, my snail could distinguish between salty, bitter, and sweet flavors. The thousands of chemoreceptor cells along its upper tentacles were similar to those inside a human nose. Snails "see" the world through smell, the way many insects do, and they can detect aromas from a few airborne molecules.

In its native habitat, my snail determined the source of a scent and the distance from which it wafted based on wind speed and direction. There was no woodsy fragrance blowing through my room, and the snail, especially while it lived in the pot of violets, must have been surprised by the kaleidoscope of unfamiliar smells, the scent of humans, human food, tea, soap, paper, and ink.

Unlike the human nose, infamous for its secretions, the noselike tentacles of a snail are the *only* mucus-free part of its body. And compared to the stationary, side-by-side nostrils of a human, the snail's two independent tentacle-noses give it a kind of stereoscopic sense of smell. I imagined a crowd of humans with smell receptors completely covering their arms, walking down the main street of a town. As they passed coffee shops, bakeries, and restaurants, their arms would wave wildly toward the aromas. Perhaps restaurant critics so endowed could, with the flourish of an arm, report not just on their own entrée but also on those of other diners at nearby tables.

Though the snail had a sophisticated method of scent tracking, I wondered how it experienced a life so devoid of sight and sound. In its native woods, my snail could not see the moss over which it glided or even the plants it climbed. It could not see the trees, nor the stars overhead. It could not hear birdsong at daybreak, nor the midnight howls of coyotes. It could not even see or hear its own kin, let alone a predator. It simply smelled and tasted and touched its world.

The closest I could come to understanding the snail's experience of its surroundings was in reading Helen Keller's portrayal of the richness of smell and touch from her own human experience in *The World I Live In:*

> I am not sure whether touch or smell tells me the most about the world. Everywhere the river of touch is joined by the brooks of odor-perception . . .
>
> . . . Touch sensations are permanent and definite. Odors deviate and are fugitive,

> changing in their shades, degrees, and location. There is something else in odor which gives me a sense of distance. I should call it horizon—the line where odor and fancy meet at the farthest limit of scent.

I wondered if my snail was aware of a scent "horizon" and how far the odor of a mushroom could float through the air. A snail's navigation is complex, based on ever-changing odors, sources of darkness and light, a tactile sense of air movement, and, through the touch receptors on its single body-foot, a response to vibrations and types of terrain. This was how my snail mapped the wild woods from which it came, as well as the crate beneath the pot of violets, and the terrarium.

I combed through scientific gastropod literature, eager to know more about my companion. I learned that snails are extremely sensitive to the ingestion of toxic substances from pollution and to changes in environmental conditions, such as temperature, moisture, wind, and vibration. I could relate to this, as my dysfunctional autonomic nervous system made me sensitive to these things as well.

Since I was unable to tolerate most drugs, my doctor prescribed treatments at such minute doses that a pharmacist said he felt as if he were dispensing medication for a mouse. My body's temperature regulation no longer worked. One moment I was chilled, and the next too hot; this made life as a cold-blooded poikilotherm seem appealing. Before my illness, I could sleep through a full moon; now even with my room pitch black at night, sleep was elusive. The sound of the telephone sent a tsunami-like shock wave coursing through me, so I kept the ringer turned off. I could listen only to music that was slow and continuous; anything with individually punctuated notes was too jarring. This restricted my entertainment to the calm of Gregorian chants at a barely audible level. I wondered if the snail could sense the vibrations through the air, and what the Benedictine monks would think of singing to a gastropod.

9. Marvelous Spirals

And seeing the snaile,
which every where doth rome,
Carrying his owne house still,
still is at home.

—JOHN DONNE,
from “To Sir Henry Wotton” (1572–1631)

Even when my snail was asleep, I loved to gaze at the beautiful spiral of its shell. It was a tiny, brilliant accomplishment of architecture, and because the radius of the spiral increases exponentially as it progresses, it fits the definition of a logarithmic or an equiangular spiral. Also known as the marvelous spiral, it accounts for the sound of the sea that one hears when an empty shell is lifted to the ear: outside noise enters the curving chamber and echoes back and forth, jumbling into a continuous surflike tone.

In 1905, G. A. Frank Knight delivered a lecture to the Perthshire Society of Natural Science in which he noted that

> the whole subject of convolution in the mollusca is one of extreme interest, and has excited the enquiries of eminent scientists . . . The animal, with unerring certainty, will mould for itself a habitation, which . . . will be finished with an absolutely perfect devotion to geometrical curves, proportion, and principles . . . How infinitely varied the series of curves may be, and how wide is the scope granted . . . [but] the law of "the spire of the logarithm" must be strictly adhered to.

In most languages, the word for "snail" refers to its spiral shape: in the Native American language Wabanaki, the term is *Wiwilimeq,* for "spiraling water creature." Giovanni Francesco Angelita, an Italian scholar, wrote an essay in 1607 titled "On the Snail and That It Should Be the Example for Human Life." He praises the creature's thoughtful pace and good morals and credits it with inspiring everything spiral, from the invention of drill bits to Europe's most famous staircases.

As a snail grows, its mantle secretes material at the shell opening, thus lengthening and widening its house by increments to keep up with its expanding body size. A snail's shell is "part and parcel of the animal itself," points out the nineteenth-century naturalist Searles Wood, as quoted in *British Conchology.* And Edgar Allan Poe, in an odd leap from his usual macabre genre, comments in the preface to *The Conchologist's First Book* in 1839 that "the *relation* of the animal and shell, with their dependence upon each other, is a radically important consideration in the examination of either."

My snail's shell had five and a half turns or whorls around its center starting point. I could see the past growth lines, and its final shell opening was elegantly rounded off with a wide, creamy lip. Was this curved lip a way to strengthen the shell edge? Perhaps it was a sort of built-in gutter system. I would learn, soon enough, that this detail proved, irrevocably, my snail's maturity.

In Italo Calvino's book *Cosmicomics,* in a story titled "The Spiral," the molluscan narrator expounds on the art of shell making and reflects on what it is like to be part shell. But it was the gastropod narrator in Elizabeth Bishop's poem "Giant Snail" that is so enchanted with its own shell that it made me want my own:

> Ah, but I know my shell is beautiful, and high, and glazed, and shining. I know it well, although I have not seen it. Its curled white lip is of the finest enamel. Inside, it is as smooth as silk, and I, I fill it to perfection.

The whorls of a snail shell lean asymmetrically out from the center. My snail's shell was dextral, with a right-side opening, as is most common. However, some snails are sinistral, with a left-side opening. In his Perthshire Society lecture, G. A. Frank Knight takes us inside for an architectural tour:

> If we imagine the interior of the shell to be a spiral staircase, then, as we ascend a dextral mollusc, the "axis" . . . of the stair would always be at our left hand, and similarly, if the mollusc be sinistral, the stair up into its interior would always curve round the axis on the right hand.

Spiral direction has an impact on relationships; a snail must find a mate of its species with a matching shell.

If a snail's shell gets injured, a repair can be made quickly. New shell material is secreted by the mantle, and where there was once a crack, a scar appears, looking much like a skin scar. Even a missing shell section can be replaced. Oliver Goldsmith described this in 1774:

> Sometimes these animals are crushed seemingly to pieces, and, to all appearance, utterly destroyed; yet still they set themselves to work, and, in a few days, mend

> all their numerous breaches . . . to the re-establishment of the ruined habitation. But all the junctures are very easily seen, for they have a fresher colour than the rest; and the whole shell, in some measure, resembles an old coat patched with new pieces.

In an article titled "Shell Fish: Their Ways and Works," published in 1852, George Johnson praises the snail's shell as "an edifice rivalling, nay exceeding, in complexity yet order of details and perfection of elaborate finish, the finest palaces ever constructed by man!" Johnson's "elaborate finish" probably referred to the colorful shiny shells of the tropics. My woodland snail's shell, however, though beautifully and perfectly formed, was an earthy color with a modest matte finish. It was better described by the Mandarin Chinese words for "humble abode," which are "wō jū" or a 蜗居, literally meaning "snail's house."

Reminding me of a rolled-up sleeping bag—the kind I had once strapped atop a backpack—my snail's shell was a brilliant solution to a life of wanderlust. And there was a further benefit. In the third or fourth century BC, the Athenian poet Philemon observed, "How ingenious an animal is a snail . . . When it falls in with a bad neighbor it takes up its house, and moves off."

• • •

Unlike the sturdy external shell of my snail, my supporting structure was internal. But the bones that made up the skeleton deep within me were losing their density at a rapid rate, and there was little I or my doctors could do to halt the problem. My status as a vertebrate was literally dissolving. I would eventually become a spineless, soft-bodied creature, more like a gastropod than a mammal. And unless my armpits could secrete shell material, I would be more slug- than snail-like.

I observed my snail's spiral shell from the outside, but what was it like to live inside such a shape? Just a month before the onset of my illness, I had visited the Guggenheim Museum in New York. Halfway down the rotunda's spiraling interior, I stopped. It was dizzying to look up as the floors curved around and above me and equally so to look down to the ground level far below. Now I tried to imagine, were I as large in proportion to the Guggenheim as the snail to its shell, what it would be like to have my head stick out the main entrance below and my body wind all the way up the spiraling floor.

10. Secret Recipes

My wide wake shines,
now it is growing dark.
I leave a lovely opalescent ribbon:
I know this.

—ELIZABETH BISHOP,
from "Giant Snail," 1969

Hundreds of millions of years ago, by chance, some marine snails evolved certain traits that allowed them to colonize land. To survive the challenges of a dry terrestrial habitat, they had to keep their bodies moist. While my mammalian ancestors evolved dry skin to prevent dehydration, my snail's gastropod clan went in a different direction, perfecting and luxuriating in the sticky thickness of slime, or mucus. While Homo sapiens have internal mucus, and more of it than we realize, it's the extravagant nature of the gastropod to be completely coated externally.

While slime can certainly be disgusting, it hadn't occurred to me that it might be interesting. So many times I had come in from gardening to wash up, dissolving the dirt instantly with water and soap, only to find that the patches of slime from inadvertent encounters with hidden slugs —the less aesthetic relatives of my elegant snail —remained impervious, sticking to my hands like glue. It took a pumice stone or even four-hundred-grit sandpaper to get the stuff off.

Slugs, despite what one might think, given their naked look, do not predate snails on the evolutionary tree but were once snails that evolved over time to be shell-less. Without a shell to tote around, they can change their shape more easily than a snail, thus squeezing into smaller crevices.

The biologists C. David Rollo and William G. Wellington once commented with amusement on their gastropod subjects: "A bag of cold water that cannot even move unless it leaks should not be able to survive outside a bog." And yet, thanks to their protective slime, terrestrial gastropods have thrived.

Slime is the sticky essence of a gastropod's soul, the medium for everything in its life: locomotion, defense, healing, courting, mating, and egg protection. Nearly one-third of my snail's daily energy went into slime production. And rather than making a single batch of "all-purpose" slime, my snail had a species-specific recipe for each of these needs and for different parts of its body. It could adjust the ingredients, just as a good cook would, to meet a particular occasion. And in a catastrophic accident in which a snail is squashed, it can release a flood of lifesaving, medicinal mucus packed with antioxidants and regenerative properties.

Skimming a chapter titled "Molecular Biomechanics of Molluscan Mucous Secretions" by the zoologist Mark Denny in *The Mollusca*, I came across a powerfully alliterative phrase that stuck in my head: "the macromolecular architecture of molluscan mucus." The technical details were beyond me, but clearly the phrase had to do with how the stuff holds itself together—how a quantity of water is controlled by a bit of salt and a protein-sugar. Denny points out admiringly that "if stirred with a rotating rod . . . [mollusk mucus] recoils when the stirring is stopped, and . . . has sufficient tensile strength to self-siphon out of a beaker."

My snail secreted a special kind of slime for locomotion, called pedal mucus, over which it traveled. While its ability to glide over a patch of moss appeared effortless, when it went up the glass side of the terrarium, I could see bands of minute ripples moving across the underside of its foot. These ripples momentarily turned the mucus from solid to liquid, disrupting friction and allowing the snail to advance at a speed of a few inches per minute. Its single-footed method of travel was far more ancient than my own bipedal ambulation or that of my quadruped dog.

"Upon this slime, as upon a kind of carpet . . . [the snail] proceeds," wrote Oliver Goldsmith. The zoologist T. H. Huxley, author of *Practical Biology*, commented in 1902 that "the wave-like contractions" of a snail's foot are "so delicately adjusted . . . that the creature can crawl with ease and comfort over a knife-edged surface."

Slime travel has intrigued some innovative researchers in the Netherlands. Hoping to improve the comfort level of colonoscopies, they are designing a small robot that can travel snail-like through the mucus-coated intestines of humans. I wondered what other snail traits might inspire further biomimicry.

Pedal mucus is an incredible adhesive; it explains my snail's ability to cross over soft mosses; proceed upside down along a leaf; or sleep, oblivious to gravity, high on the side of the glass terrarium or dangling from the tip of a fern frond. Before the snail's arrival at my bed-side, the concept of ceiling art had absorbed my attention. I would concoct in my mind various methods for safely affixing horizontal images to the white surface above me. Perhaps slime glue was the solution.

The sticking power of slime, in concert with a

muscular foot, creates a creature of Olympic caliber, as documented by E. Sandford in the *Zoologist: A Monthly Journal of Natural History* in 1886:

> **Experiments to Test the Strength of Snails**
> Perceiving a Common Snail . . . crawling up the window-blind one evening, it occurred to me that I would try [to discover] what weight it could draw after it . . . Accordingly I attached to its shell four reels of cotton which happened to lie on the table . . . I then weighed the entire load and found it to be two ounces and a quarter, while the snail itself weighed only a quarter of an ounce. Thus it was able to lift perpendicularly nine times its own weight! I then made an experiment with another and somewhat larger snail, which weighed about one-third of an ounce, the load being . . . drawn in a horizontal position on the table. Reels of cotton to the number of twelve were fastened to it, with the addition of a pair of scissors, a screwdriver, a key, and a knife, weighing altogether seventeen ounces, or fifty-one times the weight of the snail. The same snail, on being placed on the ceiling, was able to travel with four ounces suspended from its shell. I next tried it on a piece of common thread, suspended and hanging loose, with

> another snail of its own weight, which it carried up the thread with apparent ease. After this I tried it on a single horse-hair strained in a horizontal position, but it had then enough to do to crawl over this narrow bridge without a load.

Where was the Society for the Prevention of Cruelty to Animals? Apparently it was not paying close attention to snails. Perhaps the larger snail's refusal to carry a burden over the single horse-hair was due, at least in part, to sheer exhaustion after participating in so many experiments.

Snails will often reuse their own or another snail's trail in order to save on slime production. By detecting pheromones in a trail, they can determine whether it leads to foe, friend, or potential mate. Some terrestrial snails even "gallop" by picking up the front of their foot and leaping forward, leaving behind a dotted slime trail. This may save on slime use or possibly outwit a predator. If frightened, one snail species will lift itself up on its posterior and speed-glide eighteen inches per minute.

The idea of being coated from head to toe with such a slippery, sticky substance was unsettling to me. I thought that my snail would feel an equal aversion to sunning itself on a hot, sandy beach. Evolution has led to our contrasting skin anatomies and, as a result, to our opposite fears.

Part 4

THE CULTURAL LIFE

[Snails are] found to be furnished with the organs of life and sensation in tolerable perfection; they are defended with armour that is at once both light and strong; they are as active as their necessities require; and are possessed of appetites more poignant than those of [other] animals . . . In short, they are a fruitful industrious tribe . . . [They have their] . . . powers of escape and invasion; they have their pursuits and their enmities.

—OLIVER GOLDSMITH,
A History of the Earth and Animated Nature, 1774

11. Colonies of Hermits

Where'er he dwells, he dwells alone,
Except himself has chattels none,
Well satisfied to be his own
whole Treasure.

—WILLIAM COWPER,
from "The Snail," 1731

A diet of nothing but portobellos seemed monotonous, so I offered my snail a concoction of wetted-down cornstarch and cornmeal. This was the diet suggested in a pamphlet sent to me by the local Cooperative Extension office. It was a big mistake: the snail overate. It climbed in a staggering sort of way to the top of the terrarium. Clearly suffering from a severe case of indigestion, it stayed there for hours, excreting wastes from all orifices.

I was terribly worried. If the snail didn't recover from cornstarch indulgence, then how, I wondered selfishly, could I survive my illness without the snail for a companion? It was a miserable night for both of us, and I resolved never again to feed it anything unnatural. The next morning I was relieved to find the snail moving about normally, and following its usual routine, it went off to doze in a soft mossy corner of the terrarium.

A woodland snail is most at home in the soft layer of debris, leaf litter, and duff that carpets the forest. Snails are known as decomposers because they feed primarily on dead matter, thus filling a niche in the ecosystem by returning nutrients to the soil. A special enzyme allows them to digest cellulose, which explained my snail's penchant for eating paper. It is rare for native woodland snails to feed on live plants, but if they do, they usually eat older deteriorating leaves. Many species love to graze on algae and will munch contentedly on mushrooms, even those toxic to humans. The mycelium, the thread-like part of a fungus that often grows under-ground, is a favorite food.

The foraging of snails is complex; they vary their diets to balance their nutrient intake. Two snails of the same species in the same location may make different dining arrangements. They are intrigued by a new food but proceed cautiously: After inspecting it with the lower tentacles, they take a small taste. If there are no unpleasant side effects, they will return for a larger portion.

Soil is also part of a snail's diet, providing much-needed nutrients such as calcium, which is essential not only for shell growth and repair but also for egg production. So critical is this mineral that the snail is the only known land animal able to find it by smell. When my caregiver placed a little pile of crushed eggshells in the terrarium, my snail waved its tentacles as rapidly as a snail can, setting off at once to investigate and eat. From then on, the eggshell site was one of its regular hangouts.

The mussel shell of fresh water was another of the snail's favorite spots. Usually, it drank straight from the small basin, but sometimes it climbed right inside, flattening its foot into the shell's pearly contours and absorbing water directly through its skin, a hydration method known as foot drinking.

Most snail activity occurs with nightfall's cooler temperatures, or just after a daytime rainstorm, when humidity improves locomotion and replenishes the supply of fungi. Even in the house, my snail increased its activity on a rainy day. During the weeks it lived in the flowerpot it must have been perplexed when I watered the violets, as the highly localized rainstorm never resulted in any fresh vegetation or mushrooms.

In the wild, snails generally travel downwind from their daytime hiding places, then find their way home in the morning by following familiar smells. While exploring the crate, my snail had used its homing skills daily to return to the pot of violets, which had offered the only decent sleeping place. The terrarium, however, offered it numerous hiding spots, and it made use of them all.

Older snails may spend their evenings foraging within an area of several square yards, while juveniles might roam five times as far, searching, perhaps, for new food sources or their own territory in which to start a colony. Many snails live out their lives in such close proximity to where they hatch that the botanist A. D. Bradshaw once remarked, "All I can say is, you've convinced me [that this snail species] is a plant."

Colonies of snails are often confined by the natural contours and terrain of their location. A species' microenvironment might be as site specific as a hillside or a valley, or even a clump of leaves drifted against a log, or a damp area between rocks. Populations may number a hundred individuals, though they can be much smaller or, if the habitat is contiguous and extends for miles, much larger. I envisioned a colony of hermits, each off on their individual forays by night, each sequestered, asleep, by day.

When I thought of the distance my snail could travel in relation to its size, my own immobility was stark in comparison. And my life was becoming nearly as solitary as my snail's. As the months drifted by, it became harder for friends to give up weekend time to make the long drive to visit. There were many days when I saw a caregiver for just a half hour at mealtimes, and I was becoming more and more cut off from the world.

My bed was an island within the desolate sea of my room. Yet I knew that there were other people homebound from illness or injury, scattered here and there throughout rural towns and cities around the world. And as I lay there, I felt a connection to all of them. We, too, were a colony of hermits.

12. Midnight Leap

little snail
facing this way
where to now?

—KOBAYASHI ISSA (1763–1828)

I remember a summer of better health. Deep into a humid night I half woke from thirst. Keeping my eyes almost closed to maximize sleepiness, I made my way by moonlight, barefoot, across the wood floor toward the kitchen sink and water.

Suddenly I was high in the air, my body springing up of its own accord, my legs zooming into my nightgown, knees bent. I stayed suspended this way for a long moment, my mind still focused on water, unaware that it had been hijacked.

I had stepped on a slug.

Landing back on my feet, I was now fully awake, and it didn't take long to figure out why a slug was on the loose. Earlier that day, my orange coon cat, Jasione, must have taken a nap in a cool, damp garden spot. A slug had stuck in the cat's fine, soft fur, as occasionally happened. Well camouflaged, it rode its feline mount into the house at a trot. During her evening bathing ritual, Jasione must have managed to extract it from her fur. A spider she would have pounced on and eaten, but the slug, saved by its slime, was discarded alive.

As the sun set, the humidity in the house increased, improving conditions for gastropod travel. The slug made its leisurely way across the floor, unaware that I, a large mammal, was coming toward it through the dark. Though its own weight was less than a sixth of an ounce, it easily and passively repelled my one hundred pounds with its slime and, unperturbed, continued on its way.

Were I to encounter an animal as large to me as I was to the snail—Highsmith's giant carnivorous snails of Kuwa come to mind again—what would I do for defense, and how would I escape? I could not think of a single passive human defense method as brilliant as slime.

In terms of size, mammals are an anomaly, as the vast majority of the world's existing animal species are snail sized or smaller. It's almost as if, regardless of your kingdom, the smaller your size and the earlier your place on the tree of life, the more critical is your niche on earth: snails and worms create soil, and blue-green algae create oxygen; mammals seem comparatively dispensable, the result of the random path of evolution over a luxurious amount of time.

Three and a half billion years ago, when life on earth began, the snail and I shared a common ancestor, some kind of simple worm that over time evolved into two animal groups. The protostomes, which in the embryotic stage develop a mouth first and then an anus, branched off into gastropods and the species of snail at my side. And the deuterostomes, which develop the same characteristics, though somewhat embarrassingly in reverse order, anus first and then a mouth, branched off into mammals, including *Homo sapiens*.

The snail and I both had a gut and a heart and a lung, though I had two lungs to its one. There the similarities ended. Observing its quirky telescoping eye-noses, ribbonlike teeth, slimy skin, and movable house, it was hard to believe that we originated on the same planet. In 1862, Charles Darwin wrote to the geologist Charles Lyell, "I should think mammals & molluscs rather too remote from each other for fair comparison."

The evolution of a species is shaped, in part, by its unique history of viral and bacterial pathogens. By rearranging cellular DNA, pathogens can switch genes on or off and thus impact the traits of a species' future generations. Luis P. Villarreal, director of the Center for Virus Research, proposes that even common, seemingly benign viruses may have shaped human cognition and socialization. And the virologist Thierry Heidmann, as well as Villarreal, links viruses to the development of the placenta—without which we humans would still be egg layers. I wondered if there was DNA for other animal traits buried in my own genetic code. We all have some genes that for unknown reasons are in the "off" mode; perhaps scientists will someday figure out how to flip these switches, and we'll each be able to choose other interesting animal traits: a tail, striped fur, wings, or even gastropod tentacles.

And how, I wondered, did the mysterious virus that had felled me change life inside the cells of my own body? Would there ever be a switch I could flip to instantly restore my health? This was a most tantalizing idea.

As I glided further into the dusty mollusk volumes, I discovered that gastropods—which account for 80 percent of all mollusks—are one of the most successful animal groups. They have existed for half a billion years, surviving or reevolving through several mass extinction events. They make their home in nearly every habitat on earth.

While thirty-five thousand living terrestrial snail species have been documented, tens of thousands are not yet identified. The majority of these are microscopic, as Ernest Ingersoll points out in his 1881 essay "In a Snailery": "Some [snail species] are so minute that they would not hide the letter *o* in this print."

If we *Homo sapiens* thought we were in charge of the planet, here was clear evidence to the contrary. The humble snail and its clan have a far older, and stickier, foothold on the earth than we more recent creatures. It was clear to me that gastropods should make front-page headlines in the *New York Times,* and mammals, particularly the human sort, should be relegated to the back sections. But then, with its many-toothed radula, cellulose-digesting enzyme, and lack of vision, my snail was more likely to eat the *Times* than read it.

With a home range measured in yards and a locomotive speed of a few inches per minute, how did terrestrial snails colonize the world's continents? As it turned out, it was not just my cat that provided high-speed gastropod transport. Tim Pearce, a malacologist (a person who studies mollusks), specializes in gastropods. He tracked the nighttime jaunts of a group of snails by attaching a thread to each of their shells. He discovered one snail was moved, alive, twenty-seven yards by a shrew, its journey ending more than three feet below ground in a burrow.

One hundred and fifty million years ago, my snail's ancestors probably caught unexpected rides on fifty-ton dinosaurs. These largest of steeds may have provided fine dining, as fossil evidence suggests that snails enjoyed banquets of dinosaur dung. During North America's megafaunal period, which ended thirteen thousand years ago, snails may have hitchhiked on giant tree-eating sloths, elephants, and perhaps the fastest of their mounts, the lions, cheetahs, and powerful saber-toothed cats.

Yet none of these transport methods explained how terrestrial snails came to colonize islands far out to sea, and this was a problem that greatly vexed Charles Darwin. On September 28, 1856, he wrote to the naturalist Philip Gosse, "The means of transport . . . of land mollusca utterly puzzle me." A few days later, writing to his cousin, the naturalist William Fox, Darwin states, "No subject gives me so much trouble & doubt & difficulty, as means of dispersal . . . to oceanic islands.—Land Mollusca drive me mad." And clearly they did, for that December, Darwin wrote to the botanist Joseph Hooker, complaining "I have for [the] last 15 months been tormented & haunted by land mollusca."

As he later recorded in *The Origin of Species*, in 1859, Darwin wondered if a hibernating snail "might be floated in chinks of drifted timber across moderately wide arms of the sea." This set him off on his usual method of scientific inquiry: experiments. He filled containers with ocean water and obtained a collection of live, hibernating snails:

> [One snail] was put into sea-water for twenty days, and perfectly recovered. During this length of time the shell might have been carried by a marine current of average swiftness, to a distance of 660 geographical miles.

Greatly relieved by this possible explanation, Darwin confided to Joseph Hooker, "I feel as if a thousand pound weight was taken off my back," though he concludes in *The Origin of Species,* "It is, however, not at all probable that land-shells have often been thus transported; the feet of birds offer a more probable method."

Darwin's dispersal theories turned out to be right: a snail might attach itself to a bird, settling into the plumage as a stowaway on a long migratory flight. For aerial travel closer to home, a tiny snail has been known to catch a ride on the leg of a bee or may adhere to materials picked up by an avian nest maker.

Stuck to an autumn leaf, a snail may blow along in a storm, its magic carpet eventually landing in faraway terrain. It is even thought that microscopic snails may be swept up by the wind, rising on air currents until they join the fertile bank of animal and plant minutiae that inhabit the earth's atmosphere. They may float undreamed-of distances, finally descending within a rainstorm, the perfect humid landing condition for slime travel and a search for fresh fungi.

Millions of centuries of voyaging via animals, water, and wind brought my snail's family to colonize the woods near where I was staying. It was by chance that my snail's path had intersected with a human trail, just as a friend—the sort of friend who stopped for a snail—was passing by. The history of gastropod travel now included the unexpected journey of my own snail, which had arrived at my bedside by human transport.

13. A Snail's Thoughts

why
such careful consideration
snail?

—KOBAYASHI ISSA (1763–1828)

I was certain that my snail was just as aware of the details of its world as I was of mine, and so I began to wonder about its intelligence. I crept through the pages of scientific gastropod literature until I got stuck on the paragraph that describes a snail's brain, which, depending on its species, has 5,000 to 100,000 giant neurons.

A snail has memory; it can learn new smells and tastes and retain the knowledge for weeks or months, adapting its behavior accordingly. "Too many people think . . . that snails have no brains at all," writes the malacologist Ron Chase in their defense. Like humans, older snails tend to learn more slowly than younger ones. There are plenty of situations that scare a snail, and even scientists now use the term *fear* to describe a snail's reactions to danger.

In 1888, an unknown author declared in an essay titled "Snails and Their Houses" that the snail "is by no means lacking in intelligence, but exemplifies the truth of the aphorism that still waters run deep." Lorenz Oken, a German naturalist of the same century, waxed rhapsodic in his *Elements of Physiophilosophy*:

> Circumspection and foresight appear to be the thoughts of the [snail] . . . What majesty is in a creeping Snail, what reflection, what earnestness, what timidity and yet at the same time what firm confidence! Surely a Snail is an exalted symbol of mind slumbering deeply within itself.

Even contemporary malacologists seem to be aware of the complexity of an individual gastropod's life. "Clearly, to achieve any real understanding of the life of a slug or snail, the whole life history must be taken into account," explains A. J. Cain in his chapter in *The Mollusca*, "Ecology and Ecogenetics of Terrestrial Molluscan Populations." The biologists Teresa Audesirk and Gerald Audesirk note just as respectfully in their own chapter, "Behavior in Gastropod Molluscs," that "as investigators themselves learn to 'think like a snail' . . . ever more amazing feats of [snail] learning power are revealed."

An account of a snail's behavior in a tight situation intrigued me. It appeared in "Mental Powers and Instincts of Animals," a chapter in Charles Darwin's manuscript *Natural Selection*:

> Mr. W. White . . . fixed a land [snail] in a chink of rock . . . in a short time the animal protruded itself to its utmost length, & attaching its foot vertically above tried to pull the shell into a straight line; then resting for a few minutes, it stretched out its body on the right side & pulled its utmost but failed; resting again, it protruded its foot on the left side pulled with its full force & freed the shell. This exertion of force in three directions, which seems so geometrically reasoned, might have been instinctive.

Were *I* stuck in a chink of rock, I'd have tried a similar approach. This raises the unanswerable question of where instinct ends and intellect begins. My snail went about its life, moment to moment, much as I did, making decisions—or being indecisive—about food and shelter and sleep. If a snail can learn and remember, then it thinks, at least on some level; I was convinced of this. And until someone (preferably a snail) can prove otherwise, I will hold on to this belief.

The life of a snail is as full of tasty food, comfortable beds of sorts, and a mix of pleasant and not-so-pleasant adventures as that of anyone I know.

Except for their remarkable romances, about which I would soon learn, snails lead solitary lives. Their behavior is considered intermediate in complexity, simpler than that of mammals and insects but more advanced than that of worms. I wondered if they communicated with each other at all. In *The Descent of Man*, in 1871, Charles Darwin recounted the observations of a scientific colleague:

> Mr. Lonsdale informs me that he placed a pair of land-snails . . . one of which was weakly, into a small and ill-provided garden. After a short time the strong and healthy individual disappeared, and was traced by its track of slime over a wall into an adjoining well-stocked garden. Mr. Lonsdale concluded that it had deserted its sickly mate; but, after an absence of twenty-four hours, it returned, and apparently communicated the result of its successful exploration, for both then started along the same track and disappeared over the wall.

Did these two snails exchange tentacle touches? And if so, what information was relayed through physical contact, smell, and pheromones? Wouldn't a lone snail, given the option, prefer proximity to another of its species to ensure procreation and gene survival? While contemporary malacologists do not believe that snails form any permanent attachments to one another, Lonsdale's account, if true, hints at the possibility of gastropod kin selection, as a snail too ill to make its own pedal mucus could travel more easily over its companion's trail to reach food and shelter.

Aphid parents living on a leaf transmit vibratory signals to their minute offspring to warn them of predators. And though it was assumed that ants do not communicate with auditory signals, scientists have just discovered that some species utilize substrate-born sounds. Even if the world of the snail is soundless, this does not preclude other communication methods. The biologist Roman Vishniac was always amazed at the individual personalities of, and the relationships and battles between, microscopic animals in a drop of pond water. How can any species, even our own, ever fully fathom by what means another species or animal group interacts?

I respected my snail's intelligence, so it distressed me to peruse the Cooperative Extension's snail-farming literature. Historically, snails have been a healthy food source and a medicinal remedy for nearly every illness. But learning how to fatten them up—especially after the cornstarch disaster—was unsettling. I avoided glancing toward my small companion as I read, fervently hoping that it did not have any sort of gastropod telepathy and, if it did, that it understood that it was most helpful to me alive.

The Romans, however, had no such scruples; they placed live snails in Eden-like gardens of lush vegetation surrounded by sinister water-filled moats, satiating a snail's desires and preventing its escape. Still, were I a farmed snail, I'd choose ancient Rome's fresh organic produce over today's agribusiness diet of chemically raised GMO cornmeal.

Farmed snails unhappy with their lot in life have found ways to break free. In the mid-nineteenth century, Sir George Head described the single-minded survival instinct of snails for sale at a street market in Rome: "The proprietor," Sir George commented, "is obliged to exert his utmost vigilance and dexterity in order to restrain their incessant efforts to crawl over the edge of the basket and escape."

A U.S. Department of Agriculture's snail-farming bulletin notes that confined snails may form an aggregate, their combined strength and skills resulting in a breakout. I thought of hundreds of snails packed densely into shipping crates, en route to a restaurant where escargots grace the menu and boiling water awaits. With one purpose in mind, they join forces, push up with their muscular heads against the top of the crate, and pop the lid right off, gliding slowly but steadily toward freedom.

14. Deep Sleep

"I am going to withdraw from the world; nothing that happens there is any concern of mine." And the snail went into his house and puttied up the entrance.

—HANS CHRISTIAN ANDERSEN,
"The Snail and the Rosebush," 1861

A snail unhappy with its dining options or uncomfortable with the weather will go dormant. Its heart rate slows to just a few beats per minute, and its oxygen intake diminishes to one-fiftieth of its active use. Perhaps it was my insomnia combined with the way my unusable time kept evaporating, but of all the traits the snail acquired through evolution, dormancy seemed to be the very best. Like Sleeping Beauty, a snail may not wake until circumstances are favorable—though, like Rip Van Winkle, it may wake into a changed world.

During the summer months, if conditions

become too dry, windy, or hot, or if food supplies are limited, a snail will go into a kind of dormancy called estivation. It climbs up a plant, tree, or wall to be away from the earth's heat and beyond reach of predators or floodwaters. Finding a safe place, it attaches itself firmly with mucus, usually with the shell opening facing upward, which may alert it to weather changes. Then it seals up its entrance with a temporary door made of mucus. This storm door, or *epiphragm*, protects it from shifts in temperature and humidity. A snail may estivate for weeks or months, or even several years.

With winter's colder temperatures and shorter days, instead of estivating, snails will hibernate, sometimes returning each year to a familiar site. In 1835, in his treatise *On the History, Habits and Instincts of Animals*, William Kirby described prehibernation chores:

> Snails cease feeding when the first chills of autumn are felt, and . . . set about their preparations for their winter retreat . . . Each forms . . . a cavity large enough to contain its shell. The mode in which it effects this is remarkable; collecting a considerable quantity of the mucus on the sole of its foot, a portion of earth and dried leaves adheres to it, which it shakes off on one side; a second portion is again thus selected and deposited,

> and so on till it has reared around itself a kind of wall . . . It presses against the sides and renders them smooth and firm. The dome, or covering, is formed in the same way . . .
>
> . . . [Thus] it sets about erecting its winter dwelling, and employing its foot both as a shovel to make its mortar, as a hod to transport it, and [as] a trowel to spread it duly and evenly, at length finishes and covers in its snug and warm retreat.

Once in its form-fitted burrow, rather than make a single, thin epiphragm, as it would for estivation, a snail set on hibernating makes a thicker epiphragm, and depending on its species and the severity of the winter, it may make several in a row, as Ernest Ingersoll details in his essay "In a Snailery":

> Withdrawing into the shell, the animal throws across the aperture a film of slimy mucus, which hardens as tight as a miniature drum-head. As the weather becomes colder, the creature draws itself a little farther in, and makes another "epiphragm," and so on until . . . the animal [is] sleeping snugly coiled in the deepest recesses of his domicile.

These slime plates, explains the author of "Snails and Their Houses," "act on the principle of double windows, enclosing a layer of air between each pair, and so effectually protecting [the snail] from the cold."

I could not stop thinking about the making of epiphragms. Their design is specific to a snail's species and to its local climate conditions. They may be thin and simple or thick and elaborate. Strategically located breathing holes may be incorporated, or they may be permeable to air. There is quite an art to the construction of these little doors, and justly so. Despite their temporary nature, a good door in severe weather makes the difference between life and death for a snail. An epiphragm is also personal, and its statement is definitive: the snail is home but is not accepting visitors.

A combination of increasing daylight and rising temperatures will break hibernation. "The snail, having slept for so long a season, wakes one of the first fine days of April, breaks open its cell, and sallies forth to seek for nourishment," observes Oliver Goldsmith.

While many animals, including some humans, make long-distance seasonal migrations to avoid winter weather, a snail's dormancy methods allow it to stay put—a good thing, given the short distance of an average snail's excursions. The French poet Jacques Prévert wrote the "Song of the Snails on Their Way to a Funeral" about two snails that plan to attend a service for an autumn leaf that has fallen to the ground. They travel toward their destination, and when they finally arrive for the funeral, spring has come and all is happy again.

Even while dormant, some snails have continued their worldwide adventures. Here, from "Snails and Their Houses," is my favorite nineteenth-century account of relocated sleeping snails:

> Professor Morse records that he has seen certain species frozen in solid blocks of ice, which have afterwards regained their activity . . . Snails, imprisoned closely in pillboxes for two years and a half, have nevertheless survived; and a . . . snail from Egypt fixed to a tablet in the British Museum, twenty-fifth March, 1846, being immersed in tepid water, marvellously but completely recovered after an interval of four years.

I wondered what happened to snails during the last ice age, and so I asked the malacologist Tim Pearce if he thought a snail could outglide an advancing glacier. He speculated that some of the larger terrestrial snails might possibly outpace a very slow flow of ice. I thought of a tiny snail with a glacier bearing down on it. As the glacier slid closer, the temperature would drop.

In response, the snail would dig a burrow and hibernate, and the glacier would flow right over it. Even a snail couldn't last through a deep sleep of a hundred thousand years.

It came down to this: I envied my snail's many abilities. I wished I could create an epiphragm at a moment's notice and seal myself off from the challenges around me. If I couldn't, like a snail, have strength equal to many times my weight, I'd settle for just getting my normal strength back. If I couldn't glide straight up a wall or sleep stuck to the ceiling, I wished I could at least walk upright with the rest of my species. I wanted to escape from the chink of illness in which I was stuck.

How wonderful it would be if we humans with illnesses could simply go dormant while the scientific world went about its snail-paced research, and wake only when new, safe medical treatments were available. But why limit such an amazing ability to the ill? When a country faced famine, what if the entire population could go dormant to get through a hard time in a safe and peaceful way until the next growing season came around?

Part 5

Love and Mystery

Every single species of the animal kingdom challenges us with all . . . the mysteries of life.

—Karl von Frisch, *A Biologist Remembers,* 1967

15. Cryptic Life

struck by a
raindrop, snail
closes up
—YOSA BUSON (1716–1783)

My initial surprise at learning about gastropod defense systems quickly turned to respect. Whatever one's family and species, the world is thick with danger, and my snail needed all of its active and passive defenses. But the survival methods of one type of animal may seem strange to another.

Snail-eating predators come in all forms, from mammals of all sizes to amphibians, birds, and various insects, including ants, centipedes, beetles, and tinier parasites. Even a few species of spiders resort to dining on escargots, though as Simon Pollard and Robert Jackson point out in their chapter in *Natural Enemies of Terrestrial Molluscs*, venom injection by a spider requires "close contact . . . [and] tends to mean a face full of mucus, which, for most spiders, may be an unacceptable price to pay for a meal."

My snail was downright savvy; some of its active defenses were so subtle that I wasn't even aware they were strategic. Simple withdrawal into the shell not only provided physical protection but also gave the appearance that no one was home. My snail used this defense quite successfully on me the day it arrived in the pot of violets. Oliver Goldsmith notes this behavior:

> The snail, thus fitted with its box, which is light and firm, finds itself defended, in a very ample manner, from all external injury. Whenever it is invaded, it is but retiring into this fortress, and waiting patiently till the danger is over.

A snail's slow locomotive speed makes it seem vulnerable, but it may actually be a survival method, saving it from predators whose hunting activity is triggered by fast movement. The silence of its gliding also protects it from those who hunt by sound.

Being slimy is a complex defense system that goes well beyond the ability to repel a *Homo sapiens*. Large predators can't get a grip on a slippery creature, and smaller parasitic insects may get stuck in the ooze or have their mouth parts gummed up. If the usual slime recipe isn't enough of a deterrent, a special batch with particularly toxic and bad-tasting chemicals can be copiously produced on the spot. For a gastropod, survival of the fittest often means survival of the slimiest.

One well-evolved passive defense was evident in the way my snail's earth-colored shell blended into its environment. I was continually impressed by how the snail could vanish right in front of my eyes against the terrarium's flora, even when it was moving.

Then there was my snail's brilliant strategy of elusively changing its sleeping spots. It might be on its side, drawn into its shell beneath a fern frond, and thus not visible from above; or nestled against a rotting branch the color of its shell; or in a crevice, hidden by a bit of lichen. It was amazing how the snail, with virtually no sight, found such perfect hiding spots.

It was in Tony Cook's chapter in *The Biology of Terrestrial Molluscs*, titled "Behavioural Ecology," that I found the sentence that best expresses a snail's way of life: "The right thing to do is to do nothing, the place to do it is in a place of concealment and the time to do it is as often as possible."

Everything about a snail is cryptic, and it was precisely this air of mystery that first captured my interest. My own life, I realized, was becoming just as cryptic. From the severe onset of my illness and through its innumerable relapses, my place in the world has been documented more by my absence than by my presence. While close friends understood my circumstances, those who didn't know me well found my disappearance from work and social circles inexplicable.

Yet it wasn't that I had truly vanished; I was simply homebound, like a snail pulled into its shell. But being homebound in the human world is a sort of vanishing. When encountering acquaintances from the past, I sometimes see a look of astonishment cross their face, as if they think that they are seeing my ghost, for I am not expected to reappear. At times even I wonder if a ghost is what I've become.

16. Affairs of a Snail

The emotional natures of snails,
as far as love and affection are
concerned, seem to be highly developed,
and they show plainly by their actions,
when courting, the tenderness
they feel for each other.

—JAMES WEIR, *The Dawn of Reason,* 1899

One morning I looked into the terrarium and was surprised to see a cluster of eight tiny eggs. They were on the surface of the soil, just under the edge of the birch log, and were the color and size of pearl tapioca. I wondered if they were fertile and if they would hatch. I watched with interest as the snail visited the egg site every few days to tend them. On several occasions, the snail appeared to hold each egg in its mouth for a little while to "slime" it, or so I presumed, and thereby keep it at the right moisture for hatching.

Woodland snails are hermaphrodites. While rare among mammals, this characteristic is common in the majority of other animal groups and in the plant kingdom as well. A snail may find a partner randomly or show a preference for age or size. They mate in late spring, early summer, or fall, after an elaborate and complex courtship. A terrestrial snail that has been isolated for a while can, rather conveniently, self-fertilize, thus founding a new colony and ensuring the survival of its genes.

By chance, the previous year, I had watched the sensuous scene of two Burgundy snails courting in a French meadow in the film *Microcosmos*, directed by the scientists Claude Nuridsany and Marie Pérennou. Bruno Coulais' original music composition "L'amour des escargots" provides an operatic backdrop to the snails' obviously pleasurable, lengthy, lusty, and slimy embrace.

In Patricia Highsmith's short story "The Snail-Watcher," the main character observes two snails in love and is enthralled:

> Mr. Knoppert had wandered into the kitchen one evening for a bite of something before dinner, and had happened to notice that a couple of snails in the china bowl on the drainboard were behaving very oddly. Standing more or less on their tails . . . their

> faces came together in a kiss of voluptuous intensity.

Fascinated by what he's seen, Mr. Knoppert begins to read everything he can find on snails:

> [He came] across a sentence in Darwin's *Origin of Species* on a page given to gastropoda. The sentence was in French . . . [and] the word *sensualité* made him tense like a bloodhound that has suddenly found the scent.

I decided to follow in the research footsteps of Mr. Knoppert. Since he had turned to Charles Darwin for information on snail romance, so would I. My own research suggested that Mr. Knoppert may have been looking in the wrong book, as it was in *The Descent of Man* that I found the sentence, in the chapter on molluscs. It was a quote from Darwin's colleague the Swiss American zoologist Louis Agassiz. Apparently too explicit for Victorian England, Agassiz's observations had remained in the language of romance. The sentence did not contain the word *sensualité,* but it left me as curious as Mr. Knoppert, so I sent the quote off to several French-speaking friends with the resulting translation: "Whoever has had the opportunity to observe the lovemaking of snails will not question the seductiveness of their movements and airs, which anticipates the amorous embrace of these hermaphrodites."

The Victorian naturalists were eager to weigh in on a snail's love life. "The snail is, in fact, a very model lover. [It] will spend hours . . . paying attentions the most assiduous to the object of [its] affections," proclaimed the author of "Snails and Their Houses." Also smitten, the naturalist Lorenz Oken was much blunter: "Circumspection in feeling, dainty voracity, and immoderate lust appear to constitute the spiritual character of the . . . Snails."

And then William Kirby mentioned something that sounded implausible: A snail's "courtship is singular, and realizes the Pagan fable of Cupid's arrows, for, previous to their union, each snail throws a winged dart or arrow at its partner." I read more about these curious darts in Gerald Durrell's autobiography *Birds, Beasts, and Relatives.* Durrell was ten years old and living with his family on the Greek island of Corfu when he happened into a forest just after a rainstorm: "On a myrtle branch there were two fat, honey- and amber-coloured snails gliding smoothly towards each other, their horns waving provocatively." Durrell is intrigued:

> As I watched them they glided up to each other until their horns touched. Then they

> paused and gazed long and earnestly into each other's eyes. One of them then shifted his position slightly so that he could glide alongside the other one. When he was alongside, something happened that made me doubt the evidence of my own eyes. From his side, and almost simultaneously from the side of the other snail, there shot what appeared to be two minute, fragile white darts . . . The dart from snail one pierced the side of snail two and disappeared, and the dart from snail two performed a similar function on snail one . . . Peering at them so closely that my nose was almost touching them . . . [I watched as] presently their bodies were pressed tightly together. I knew they must be mating, but their bodies had become so amalgamated that I could not see the precise nature of the act. They stayed rapturously side by side . . . and then, without so much as a nod or a thank you, they glided away in opposite directions.

The "love darts" Durrell describes are tiny, beautifully made arrows of calcium carbonate, and they look as if they've been crafted by the very finest of artisans. They are formed inside the body of the snail over the course of a week and can be as much as one-third the length of the shell. The dart's shaft is hollow and circular and, depending on species, may have four finlike blades, which are sometimes flanged; one end is harpoon sharp, while the other end comes to a flair with a corona-like base.

Some species produce a new dart for each mating; others withdraw and reuse them in successive matings. A particular species might keep just one dart in stock; others have a "pouch" with a pair or more. In *Practical Biology*, T. H. Huxley comments on these Cupid's darts: "In the *spiculum amoris* . . . we have a structure, almost without parallel in the whole animal kingdom."

The trauma of being hit by a dart, however, can sometimes put a snail off its courtship. Darts are not technically necessary for mating, and less than a third of all snail species are dart shooters. It is thought that the dart transmits a slime containing special pheromones that may improve the safe storage of the partner's sperm.

A romantic encounter between a pair of snails can take up to seven hours from start to finish and involves three phases. First there is the lengthy courtship, in which the snails draw slowly closer, often circling each other, smooching, and exchanging tentacle touches. If they find they are not quite to each other's liking, they may end their romance, but if things are proceeding well, then in some species, dart shooting occurs.

In the second phase, the snails embrace in a spiral direction and mate. Some species of snails simultaneously swap sperm, while others will be male or female at a particular mating and then reverse their roles the next time. Apparently being a hermaphrodite is not always easy; if two snails of a species that take on gender roles want to be the same gender simultaneously, a conflict may occur. Regardless of the method, and assuming all goes well, sperm is exchanged either internally or externally; depending on the snail species, it may be offered in elaborately designed and decorated packages called spermatophores.

Consummation is followed by the last phase, resting; the snails, still quite near each other, both withdraw into their shells and remain immobile, sometimes for several hours. Regardless of the mating methods of a particular species, fertilization occurs internally, after the lovers have parted.

In Highsmith's story "The Snail-Watcher," I could now understand why Mr. Knoppert's wife "squirmed with embarrassment" when he "narrated snail biology to fascinated, more often shocked friends and guests." Even Durrell is so surprised by what he sees that he consults his mentor, the biologist and zoologist Theodore Stephanides. Durrell's brother Lawrence, previously bored with discussions of natural history, suddenly becomes quite interested:

"Good God," cried Larry. "I think it's unfair. All those damned slimy things wandering about seducing each other like mad all over the bushes, and having the pleasures of both sensations. Why couldn't such a gift be given to the human race? That's what I want to know."

"Aha, yes. But then you would have to lay eggs," Theodore pointed out.

"True," said Larry, "but what a marvellous way of getting out of cocktail parties—'I'm terribly sorry I can't come,' you would say. 'I've got to sit on my eggs.' "

Theodore gave a little snort of laughter.

"But snails don't sit on their eggs," he explained. "They bury them in damp earth and leave them."

"The ideal way of bringing up a family," said Mother, unexpectedly but with immense conviction. "I wish I'd been able to bury you all in some damp earth and leave you."

Gerald's mother may have been impressed with another perk of snail parenting: a snail can keep its partner's sperm alive for several months—even up to several years, if necessary—waiting for the best environmental conditions before proceeding to fertilize and then lay its eggs. My snail had probably encountered a romantic partner either very early in the spring or sometime during the prior year. The lack of predators and the provisions of large portobellos and a steady water supply were just the encouragement a prospective snail parent needed to go ahead with egg laying.

Eggs are usually laid below ground in several clutches of thirty to fifty each. My snail may have laid so few eggs and kept them above ground because the conditions in the terrarium were slightly too wet that week. Burying the eggs in such a circumstance might have been unsafe, since they could have burst as a result of osmosis.

As the embryonic snail grows, it absorbs some of the calcium from its protective eggshell. On hatching, it will eat whatever remains of the shell, and if food sources are scarce, it may also eat a nearby unhatched egg or two that would otherwise have been a sibling.

17. Bereft

the snail
has vanished! where it's gone
nobody knows
—YOSA BUSON (1716–1783)

One morning I searched for the snail, but as usual it was hard to find. I looked again among the ferns and mosses and around some lichened branches. It was not foraging for calcium at the pile of crumbled eggshell. It was not by the little tree, nor was it near the mushroom. It was not high on the terrarium glass, nor was it by the mussel shell. It was not by the little batch of eggs it had laid several weeks earlier. It was not in any of its many hiding places. It had vanished.

There was no glass top on the terrarium. Since it was the home of a living, breathing creature, I thought ventilation might be important. As far as I knew, the snail had never before left the terrarium. Even while sleeping in the pot of violets, it had always returned from its farthest expeditions.

Now, inexplicably, it was gone. Perhaps, with its eggs laid, it was finally determined to head back to its wild woods. It was probably as homesick as I was. But I simply couldn't fathom my existence without it. Its tiny sleeping presence had comforted me by day, and its explorations had entertained me by night.

I wondered if I could find and follow its slime trail, but the dry wood of the crate left no trace, and I was too weak to get down on the floor and search for further clues. From my bed I dropped pieces of mushroom onto the floor, hoping that the snail would appear. There were endless places in the room where it could hide—it could be anywhere—and I feared someone might step on it. I dreaded the sound of a terrible crunch.

As the hours passed, the situation seemed more and more futile, and I realized that I was almost more attached to the snail than to my own tenuous life.

There is a certain depth of illness that is piercing in its isolation; the only rule of existence is uncertainty, and the only movement is the passage of time. One cannot bear to live through another loss of function, and sometimes friends and family cannot bear to watch. An unspoken, unbridgeable divide may widen. Even if you are still who you were, you cannot actually fully be who you are. Sometimes the people you know well withdraw, and then even the person you know as yourself begins to change.

There were times when I wished that my viral invader had claimed me completely. How much better to live an exuberant life and then leave as one exits a party, simply opening a door and stepping out. Instead, the virus took me to the edge of life and then left me trapped in its pernicious shadow, with symptoms that, barely tolerable one day, became too severe the next, and with the unjustness of unexpected relapses that, overnight, erased years of gradual improvement.

In a March 2009 article in the *New Yorker*, Atul Gawande wrote, "All human beings experience isolation as torture." Illness isolates; the isolated become invisible; the invisible become forgotten. But the snail . . . the snail kept my spirit from evaporating. Between the two of us, we were a society all our own, and that kept isolation at bay. The snail was missing, and as the day waned, I was bereft.

18. Offspring

[The snail] drops a cluster of thirty to fifty eggs looking like homeopathic pills . . . Under the microscope the translucent egg-envelopes present a beautiful appearance, being studded with glistening crystals of lime, so that the infant within seems to wear a gown embroidered with diamonds.

—ERNEST INGERSOLL, "In a Snailery," 1881

That evening I was expecting a friend who had traveled a long way to see me. But all I could think about was the missing snail. When my friend arrived, she looked into the terrarium and lifted up a piece of moss. There, in a hole it had dug, was the snail, along with another, much larger clutch of eggs.

I had allowed the terrarium to dry out just a bit and its condition was now more favorable for egg laying. Thus the snail had burrowed under the moss and deposited its eggs where they would be well hidden and stay evenly moist. The terrarium was an expectant snail's dream, a safe nursery for hatching offspring.

My snail had recognized and dealt competently with the changing humidity, which it continued to monitor—periodically tending the eggs laid on the surface, but visiting the buried eggs only a couple of times. Though why assume that a gastropod would be any less skilled at planning for offspring than a *Homo sapiens*?

Eventually I would learn that I may be the first person to have recorded observations of a snail tending its eggs. Malacologists would have guessed that a snail visiting its eggs was more likely to eat them than to provide care. Because the first clutch was laid on the surface of the soil and numbered so few, I could see that none of the eggs were missing after the snail's visits. In the wild, revisiting eggs could give a predator a fresh trail to follow, but my snail was free from those concerns. Since it was separated from its colony, the survival of its genes was critical; perhaps this had triggered more attentive egg care.

While too much moisture can endanger eggs, they can withstand surprisingly dry conditions. "The vitality of snails' eggs almost passes belief," says Ernest Ingersoll:

> They have been so completely dried as to be friable between the fingers, and desiccated in a furnace until reduced to almost invisible minuteness, yet always have regained their original bulk upon exposure to damp, and the young have been developed with the same success.

As a result of so much egg laying, my snail lost a noticeable amount of weight; its whole body shrank in comparison to its shell size. For about a week it spent more time than usual sleeping, and then it began to eat mushroom ravenously.

I never saw the first clutch of eggs hatch. This probably occurred at night, and in addition to my flashlight, I would have needed a magnifying glass. One morning I noticed that some of the original eggs had disappeared, and when I looked closer, I saw a few tiny snails moving around; if they hadn't been moving, I wouldn't have detected them. "The young one[s] emerge in a lovely bubble-like shell," wrote the author of "Snails and Their Houses." Their shells are translucent and "so delicate," William Kirby notes, "that a sun-stroke destroys them."

The hatchlings liked to hang out on the underside of the mussel shell, probably because of the moisture, darkness, and available calcium.

Sometimes they would sleep beneath a slab of portobello, where they were out of view until they climbed up for breakfast in the evening and then were noticeable against the mushroom's white flesh. The number of hatchlings increased as the weeks passed, and I realized that additional clutches of eggs must have been laid. Perhaps the snail had deposited them at the site of the original buried group, since it revisited that site several times, though I couldn't see precisely what was happening. Or there may have been other buried egg sites.

As the tiny snails grew, their shells increased in size and slowly became opaque. There must have been several weeks between hatchings, as it was easy to tell the clutches apart. One night, a younger hatchling followed one of its older siblings across the terrarium's glass side. It then crawled onto the older sibling's shell. The older sibling turned and looked at the younger one, and they waved their tentacle-noses wildly at each other, but there was no way for the older snail to get the youngster off its back. It seemed to be a case of sibling conflict. I didn't want to interfere, but I finally managed to sit up just long enough to detach the smaller snail and place it by the pile of crushed eggshells. It spent the evening there, eating contentedly, which made me think perhaps it was after the calcium in the older sibling's shell.

• • •

I wondered how soon the little snails would mature, and I watched them closely. The thought of ending up with some hundred or so fertile snails was a bit mind boggling; it was an outcome best avoided. Highsmith's story "The Snail-Watcher" opens with one of her foreboding first lines: "When Mr. Peter Knoppert began to make a hobby of snail-watching, he had no idea that his handful of specimens would become hundreds in no time."

While the bathroom habits of my original snail had not been bothersome—a small, neat squiggle now and then on the mussel shell or terrarium glass—the casts of so many at once, especially with their fast rate of growth, was leading to a rather splotched look everywhere.

Given its solitary nature, I wondered how my snail was coping with a population explosion of its own creation. In the wild, nearly half an egg clutch is lost to weather, predators, or hungry first-hatched siblings, but in the terrarium the outcome was far more successful. I could only guess at the total number of offspring, as they were impossible to count; by day, each one had its own hiding place, and at night they were out and about, moving around in all directions at once. While watching my solitary snail had been peaceful and calming, watching a plethora of its young in simultaneous motion was hypnotic. I had to admit that *I* was just a bit overwhelmed.

Over several months, there was a gradual improvement in my condition—not so much that it was noticeable day to day, or even week to week, but I could now sit in a chair for a few minutes a couple of times a day. I wanted to try moving home, though I wasn't certain I'd be able to manage with less help. Since the prospect was daunting, I decided to leave the original snail and one of its offspring with my caregiver. Several friends, intrigued by my enthusiastic "snail reports," eagerly adopted a few of the offspring as well. The rest of the numerous progeny were released into the wild where their parent snail had been found. It was only then that an official count was made: 118 offspring had hatched.

Part 6

Familiar Territory

The crucial first step to survival in all organisms is habitat selection. If you get to the right place, everything else is likely to be easier.
—EDWARD O. WILSON, *Biophilia,* 1984

19. Release

Climb Mount Fuji
O snail
but slowly, slowly
—KOBAYASHI ISSA (1763–1828)

By midsummer, my dog, Brandy, and I were moved home. It was hard to say which of us was happier. Her cedar bed was in its familiar place, positioned in the living room to catch the morning sun. From my own bed in this same room, there was so much to take in that it was hard to know where to look first. There were the sturdy posts and beams that framed the space around me; the art on the walls, so full of color and life; and the window at my bedside, with its view of the natural world.

In the middle of the night, I was sometimes startled awake by an always mysterious bang coming from somewhere upstairs, but I felt only amused fondness for the escapades of the resident centuries-old ghost. I was used to the familiar eccentricities of my house, and this eased the transition of the move, though the adjustment to less day help was difficult.

I missed the companionship of my original snail, but the time had come to return it to its wild woods. I hoped that by fall I'd be managing well enough that its single remaining offspring could come to stay with me for the winter.

Snails with the longest life spans are often found in the most rugged climates. Given New England's deep winters, my snail would probably live several more years. There would be further lengthy courtships and additional generations of offspring. After its sheltered life in the terrarium the snail would have to readjust to the challenges of the woods, the dangerous predators, and the unpredictable weather. But with its many methods of defense and its dormancy skills, it had survived before, and I felt certain that it would do so again.

I wished I could attend the snail's release, but now that I was home, I was too far away. A letter arrived from my previous caregiver, describing how she had left the one remaining offspring in the terrarium and carried the original snail back to the place in the woods where it had been found:

On a misty day I took the snail out to a spot beneath an old oak tree. I set it on top of a wild mushroom. The snail became interested in the situation. It came partway out of its shell and then extended its head out over thin air, gradually moving its body downward, until it touched ground while still having its tail up on the mushroom's cap. Gracefully, it brought down its shell and tail, and with its tentacles pointed straight ahead, it made steady progress over leaves and twigs for the shelter of a downed oak limb.

The original snail and I had been fellow captives, but now we had both returned to our natural habitats. As I tried to make my life livable within a few rooms of my house, I wondered how the snail was coping in its native woods. Though I was home, I was still not free from the boundaries of my illness. I thought of the terrarium's limited space, and how the snail had seemed content as it ate, explored, and fulfilled a life cycle. This gave me hope that perhaps I, too, could still fulfill dreams, even if they were changed dreams.

Being home again was the next best thing to a cure, and though my physical limitations were still great, I was no longer completely bedridden. I was able to make occasional, brief but satisfying journeys within the house. I might retrieve some papers from a few yards away in the late morning, and then in late afternoon I'd try a rash trip around the corner to the kitchen for a fresh glass of water. I was elated to be able to manage these tiny tasks, though I paid dearly in exacerbated symptoms.

From my bedside window I could follow the ever-changing weather—the wind's gentle stirrings and rages, the varied moods of rain, the interplay of sun and moon and clouds. And in the midsummer heat, the gardens surrounding my farmhouse were alive with color.

There was the constant activity of small creatures flying among my perennials: hummingbirds and butterflies, moths, wasps, bumblebees, and countless other insects. There were so many different flight patterns, and the variety of wing shapes, body sizes and structures, and types of landing gear was impressive. The flow of aerial activity was so dense that I thought of it as a miniature version of New York's La Guardia Airport. Given the chaos of different species whooshing by all at the same time, it was astonishing that there weren't constant collisions.

As I window-watched, I observed the comings and goings of my neighbors; they, too, were part of the rhythm of my familiar rural landscape. They would depart for work or errands and later return, walk their dogs, cut firewood, and check their roadside mailboxes. As twilight deepened, the low dart of a nighthawk over the field would catch my eye. Darkness brought the sparking of secret codes from the mate-seeking fireflies. Then, black on black, the swift shapes of bats would swoop for late-night morsels, and the hooting of owls would come softly, softly, from the woods—until all was quiet and still beneath the ancient brightness of distant stars and the shape-shifting moon.

20. Winter Snail

closing the door
he drops off to sleep
snail
—KOBAYASHI ISSA (1763–1828)

The months passed, and leaves of flaming reds and oranges floated past my window, scattering and drifting. I was well settled at home, so the snail's single remaining offspring came to keep me company. This time the terrarium was a huge antique glass bowl with a forty-eight-inch circumference—it created a wonderful green spherical world. The juvenile snail was about one-third the size of an acorn. It took to sleeping during the day inside a hollow, rotting birch branch, which provided a perfect dark, damp hiding place. Occasionally I'd use a flashlight to look in and check on it.

As the days grew short, the winter's stillness was broken by the abstract, shifting patterns of snow in air. I watched the flakes change their shape and size, moment to moment, as they played on the wind. They would rush downward, only to rise on an updraft, swirl gracefully around, and then descend once more, vanishing into the whiteness that surrounded the house. Periodically, howling blizzard winds would hide the dark green spruce woods from my view and leave behind an even thicker snow covering.

Beneath this cold blanket, wild snails were hibernating, snug in their burrows. Did they dream while they slept, and if so, were the dreams composed entirely of smell and taste and touch? Or had they drifted into a sleep so deep as to be without thought or memory?

Inside my house the weather conditions were quite different. The oil furnace kept the air warm and dry. Instead of digging a hole and hibernating, my young snail estivated for several weeks at a time; it would either disappear inside the hollow birch branch or dangle upside down from the underside of a polypody fern. When it woke, it would eat mushroom and soil, drink water, and rasp at the inside of the mussel shell for minerals. Then it would head for the dark birch hollow or climb back up a fern frond and estivate again.

There was a paradox of speed in relation to distance and time that began to intrigue me: in contrast to its slow locomotion, my snail's life cycle was quick. In seventy years it could produce seventy generations, compared to the three generations a human might produce. Although the snail moved more slowly than a human through the physical world, it traveled more quickly than a human along its pathway as an evolving species.

My snail's quick life cycle also brought to mind a paradox within my own human world. While some aspects of society—such as technology and communications—were continually speeding up, other aspects, such as health care, moved at a pace even slower than my snail. During the months that I waited for appointments, underwent tests, and tried new treatments, my original snail laid its eggs, hatched its young, was returned to the woods, and then, in late fall, went into hibernation.

As the winter months passed, I noticed a change in my snail-watching behavior. The previous spring, when I could do almost nothing, spending time with a snail had been pure entertainment. But as my functional abilities improved just a bit, watching a snail began to take patience. I wondered at what point in my convalescence I might leave the snail's world behind.

The original snail held a place in my heart forever, and while I was fond of its offspring, it was often estivating, and I was often distracted by other things. Friends would stop by and take Brandy out for a winter hike on the woods' trails. From the window I'd watch as my dog bounded through the snowdrifts. For sheer pleasure, she'd nosedive into their depths, rolling onto her back for an icy bath, her paws waving ecstatically at the winter sky.

Neighbors checked in on me, bringing the recent local news: a cow had taken off and was eventually found wandering through the woods, and folks out skiing behind their own house on an unseasonably warm late-February afternoon had popped off their skis, stripped down to their undergarments, climbed up a large boulder covered with leafless vines, and sunbathed, only to find days later that they were unexpectedly itchy—a rare outbreak of midwinter poison ivy. And there were stories I'd missed while I was away: a neighbor's dog had arrived home one spring day with a wild turkey egg, unharmed, held gently in his mouth.

I was grateful for, and appreciative of, my closest friends and neighbors, but I still missed the outer layers of society—the acquaintances with their dual air of familiarity and mystery, and the interesting newcomers, who enliven everything. Each relapse contracts my life down to the core. And each time I've started to make my slow way back, over many years, toward the life I once knew, I find that nothing is quite as I remember; in my absence, the world has moved forward.

The banks of snow were melting away, and the air hinted of the coming spring.

The young snail was still estivating on a fern frond. Thinking that it would be hungry when it awoke, I put a fresh mushroom in the terrarium and wondered if the snail would sense the lengthening days. I was eager for open windows and the chance to be outside, even if just a few feet from the house. I wrote to one of my doctors:

> *I could never have guessed what would get me through this past year—a woodland snail and its offspring; I honestly don't think I would have made it otherwise. Watching another creature go about its life . . . somehow gave me, the watcher, purpose too. If life mattered to the snail and the snail mattered to me, it meant something in my life mattered, so I kept on . . . Snails may seem like tiny, even insignificant things compared to the wars going on around the world or a million other human problems, but they may well outlive our own species.*

21. Spring Rain

in this falling rain
where are you off to
snail?

—KOBAYASHI ISSA (1763–1828)

The first cold rains arrived, and as the weeks passed and the weather warmed, the spring peepers and wood frogs began to sing in the evenings. With the increased humidity, the young snail woke up, climbed down from its fern frond, and became active again. It would soon be mature, and the time had come to release it to the wild so that it could establish a territory and search for a mate. Life without a snail was hard to contemplate. I would miss its quiet presence, but I knew the spring showers would ensure an ample supply of fresh food, giving it the best chance of survival.

I wrote another letter to my doctor:

We have rain again today. I've been looking out the window from my daybed, wishing I could do what I'd do if it weren't for this illness, which is to put on my boots and raincoat, grab a shovel, and move dozens of plants around. One spring, in a rain shower, I dug up all my tulips in full bloom and wandered around the yard holding them by their two-foot necks, with the bulb and roots dangling down and the tulip flowers staring up at me with their big Cyclops-like eyes. I decided, based on color, just where to relocate each one. If you move plants in the rain, they hardly even know it, and they did just fine. Today is a perfect snail-letting-go day.

Hatched and raised in a terrarium, the young snail had been served the finest of portobello mushrooms and fresh water in a blue mussel shell. It had never encountered the dangers that lurked in the woods. It would have to survive on its own ingenuity, and I hoped that it would find its new home an interesting and delicious place, both familiar and very surprising.

I could now occasionally manage to walk the short distance to the edge of the woods. One evening, after a light rain had turned to drizzle, I carried the young snail to a spot beneath some large hardwood trees backed by a stone wall.

Placing it gently on the ground, I watched as it came partway out of its shell. Its tentacles lengthened, twitching with interest, and it moved them this way and that in response to the abundance of fresh odors. It explored some dead leaves, some dark green moss, a bit of lichen, and the big root of a tree. I watched it glide slowly through the dusk and vanish into the dark.

For the first time, the young snail was in a world without boundaries. I wondered what it would think of this unexpected freedom. What kind of nighttime adventures would it have while I slept, and where would it hide the next morning for its daytime rest? How would it choose a territory in an endless wilderness?

22. Night Stars

Humanity is exalted not because we are so far above other living creatures, but because knowing them well elevates the very concept of life.

—EDWARD O. WILSON, *Biophilia,* 1984

My gardens were awakening, and whenever possible I was outside on a chaise longue with Brandy at my side. We watched the sunlight find its way through the branches of the crab apple tree, dappling the blue squill and crocus, and we looked for the pointed noses of tulip leaves as they emerged in the perennial beds. Each week more flowers came into bloom, and the hedge that bordered the garden began to fill with nesting birds. Having flown several thousand miles, the ruby-throated humming-birds appeared and took up their usual summer residence in the old apple trees. They spent their time zooming between the flower beds in front of the house and the patch of poppies in the back, competing for nectar in midair with the multicolored butterflies in an ancient interspecies dance.

I could close my eyes and feel the sun warm my whole length and the wind ruffle its way over me. My ears filled with the dozy hum of bees and those tiny and odd insect sounds that rise up all around, the sounds mingling in my mind with the good, deep smell of earthy life.

Spring turned to summer, summer turned to fall, the snow came, and the snail and its offspring were still much in my thoughts. The original snail had been the best of companions; it never asked me questions I couldn't answer, nor did it have expectations I couldn't fulfill. I had watched it adapt to changed circumstances and persevere. Naturally solitary and slow paced, it had entertained and taught me, and was beautiful to watch as it glided silently along, leading me through a dark time into a world beyond that of my own species. The snail had been a true mentor; its tiny existence had sustained me.

Late one winter night I wrote in my journal:

> *A last look at the stars and then to sleep. Lots to do at whatever pace I can go. I must remember the snail. Always remember the snail.*

epilogue

Perhaps then,
someday far in the future,
you will gradually,
without even noticing it,
live your way into the answer.
—RAINER MARIA RILKE, 1903,
from *Letters to a Young Poet,* 1927

My snail observations are from a single year of nearly two decades of illness. I have merged that story and a few nonsnail stories with my later scientific readings. The research for this book and the gradual process of its writing matched the slow pace of its protagonist and were just as nocturnal. I found myself, once again, following the snail deep into its life.

While I was snail watching, there was much I did not know about my small companion, and there was just as much I did not know about my illness. I was curious about my snail's species, and solving that puzzle would take several attempts and the help of a few experts. Even more challenging was the mystery of the pathogen that had forever changed the course of my life, and I would track down the likely culprit. There was also the unknown future—my own, and that of all living things.

A Question of Species

The snail and its offspring were wild creatures. They represented half a billion years of gastropod evolution. I wanted to understand their place in such a venerable lineage.

From a book by John Burch, *How to Know the Eastern Land Snails*, I learned that my snail was of the subclass Pulmonata, having a lung and making temporary epiphragms for dormancy, instead of the permanent operculum that is attached to the foot of snails of some species, allowing them to shut their door each time they withdraw into their shell.

There are sixty families of Pulmonata land snails in the world, encompassing some twenty thousand species, and so I continued on, discovering its order—Stylommatophora (eyes at the tips of the tentacles, mostly terrestrial)—and family—Polygyridae (reflected shell lip and large size).

As to genus and species, I was at a standstill. It would take an expert to make that final determination, based on information I didn't have, such as whether the interior of the shell had a toothlike "nub," which I couldn't have seen with a live snail inside.

I contacted Tim Pearce, assistant curator and head of the Section of Mollusks at the Carnegie Museum of Natural History, as well as the biologist Ken Hotopp at Appalachian Conservation Biology. In a series of e-mails, Tim and Ken discussed the identifying details they could glean from my photos. They considered shell depth and number of whorls and even the color of the eye granules at the tips of the tentacles, until finally an agreement was reached on the snail's genus and species: *Neohelix albolabris. Neo* for new, *helix* for spiral, and *albolabris*, meaning white-lipped.

"White-lipped forest snail" is the common name, and these North American natives roam the humid woodlands as far south as Georgia, as far north as Ontario and Quebec, and west to the Mississippi.

Invisible Boundaries

The earth is home to millions of potential pathogens, of which a thousand or so depend on human hosts. The pathogen I contracted was, in its own way, an author; it rewrote the instructions followed within every cell in my body, and in doing so, it rewrote my life, making off with nearly all my plans for the future.

My illness had started with flulike symptoms and some paralyzed skeletal muscles. Within weeks it had turned into systemic paralysis-like weakness with life-threatening complications. After a slow and partial recovery over three years, I had successive severe relapses. Specialized testing brought a diagnosis of autoimmune dysautonomia, a dysfunction of the autonomic nervous system, which can cause paralysis of the circulatory and gastrointestinal systems.

Dysautonomia can make it difficult for a person to stand or sit upright because the blood vessels cannot maintain circulation against the pull of gravity. Astronauts experience this problem when readjusting to the earth's gravitational field. At one end of the "orthostatic intolerance" spectrum is syncope: a person might stand up but then immediately faint. At the other end of the spectrum are cases like mine: in an upright position, the body just gets weaker and weaker as it tries unsuccessfully to maintain blood pressure. The ability to be upright is a recent evolutionary adaptation, and it is still surprisingly fragile. The weight of the world doesn't pin me down figuratively; it pins me down literally. Horizontal surfaces are forever my flotation cushions through life.

I was also diagnosed with chronic fatigue syndrome (CFS), also known as myalgic encephalomyelitis (ME), a badly named, severe, postinfectious condition that involves permanently reduced blood volume, autonomic disorders, and genes that have been deactivated.

Seven years into my illness, further testing would reveal a clearer diagnosis: I had an acquired mitochondrial disease. Mitochondria are the "powerhouses" in each of the cells in our bodies, and they are found most densely in skeletal and autonomic muscle tissue. They metabolize nutrients and oxygen into energy in an intricate two-hundred-step process. Each of us is born with a number of unique genetic mutations, and we "acquire" additional mutations throughout our lives. A particular mutation, "unmasked" by a particular pathogen, may result in a mitochondrial error, which can then cause metabolic disease.

In my case, the pathogen may have been the virus that was spreading through the small European town I visited. Or it may have been something in the hotel water I drank one night. There was also the sick surgeon next to me on my flight home, though by then I had already succumbed to strange and severe symptoms.

Fifteen years into my illness, I would learn about tick-borne encephalitis (TBE), a member of the Flaviviridae family, which includes West Nile virus. Lyme disease can be a coinfection with TBE, though if so in my case, it resolved. TBE is not yet known to have crossed the Atlantic to North America, and my United States doctors, at the time, would not have recognized the symptoms. But its bizarre, biphasic onset matched my illness presentation: flulike symptoms, followed several weeks later by systemic paralysis-like weakness and autonomic dysfunction, with a poor long-term prognosis.

Coda

Pathogens, those critical ingredients in the primordial ocean from which life originally emerged, helped shape all species, and it was because of a pathogen that I had found myself nose-to-tentacle with a snail.

While illness keeps me always aware of my mortality, I realize that what matters most is not that I survive, nor even that my species survives, but that life itself continues to evolve. As the Holocene mass extinction rushes on, which species will be left? And what new creatures will evolve that we cannot now imagine—for what creature could ever have imagined us?

At the moment, we humans are lucky to coinhabit the earth with mollusks, even if we are a recent presence in their much longer history. I hope the terrestrial snails, secreted away in their burrows by day across the earth's vast landscapes, will continue their mysterious lives, gliding slowly and gracefully through the night, millions of years into the future.

Acknowledgments

Book writing is mostly a solitary pursuit, but this particular project brought me partway out of my shell. If it weren't for E. LaRoche, I might never have written the earlier essay that led to this book. I owe thanks to E. Somers at the *Missouri Review,* as well as to C. Mason, both of whom noticed the snail story when it was barely hatched. When I began to read about gastropods, I did not know I would fall in love with malacology literature. As I wove thousands of scientific words into my personal snail observations, M. Porter's excellent developmental editing skills, particularly her fearless and flawless insight as to what to cut, were invaluable. Undaunted by my endless revisions, M. took the project under her wing, reading every draft and surviving the feat miraculously.

Great appreciation and respect go to L. Osterbrock, D. Dwyer, and P. Blanchard; each of them brought editorial savvy to my pages and responded with good humor even when hearing from me out of the blue and sometimes at unusual hours; it is a lucky writer who has editorial friends with perfectionist streaks in different time zones. When I thought I had finished the manuscript, J. Babb smartly nudged me into doing one last editorial pass, which turned out to be essential. Thanks to L. Babb for her response on a pivotal chapter.

The following friends read one or more drafts, and their terrific questions, thoughts, and suggestions helped me shape and deepen the story: K. Adams, D. Smith, A. Levine, D. Graham, D. R. Warren, P. Kamin, L. Fisher, and S. Lester. Astute advice and support came at various junctures from J. Hamilton, T. Coburn, and J. Babb. Thanks to the MacDowell Colony and the Vermont Studio Center, and my heartfelt gratitude to S. Tullberg for making impossible dreams happen.

Timothy A. Pearce, who must have been a gastropod in his past life, is a remarkable malacologist. He answered zillions of questions with astonishing patience, thoughtfulness, curiosity, and infinite knowledge. Every time I glided too far into gastropod territory and got stuck, Tim came to my rescue. Great appreciation to the biologist Ken Hotopp, who knows just where to find a New England *Neohelix albolabris* and exactly what it might be up to at any given moment. I was lucky to have Tim and Ken as snail consultants. The thought-provoking, occasionally startling, and sometimes funny conversations and correspondence I had with them expanded my understanding of these small animals and their place in the world. If any error in malacology information has found its way into these pages, it is most certainly mine, not theirs.

Gratitude to the wetlands ecologist A. Calhoun for her early avid reading of the manuscript; to K. Vencile for his fascinating feedback; to Dr. R. Smith for his infectious-disease knowledge and his interest in malacology; to the staff at the University of Maine Cooperative Extension; and to the wonderfully kind and always helpful librarians at my local library.

I am also indebted to the nineteenth-century naturalists whose words enrich these pages. They observed every nuance of snail behavior, and their lyrical writings are unconstrained by today's more technical scientific language.

A special thanks to N. Glassman, who found the snail and without whom this story would never have happened. Appreciation to H. Schuman for sharing his love of words and to J. Miles for sharing her love of the natural world. For fulfilling a lifelong longing for island writing, I owe thanks to the Websters. To Kathryn Davis: you gave me the gift of my own words—few gifts are as great.

My agent Ellen Levine and my editor Elisabeth Scharlatt believed in a very small story about an even smaller creature, and despite the hurried pace of the publishing field, they had the patience to wait for the final draft. Thanks, also, to the good staff at Algonquin Books and Workman Publishing; to R. Careau for her excellent, meticulous, and thoughtful copy-editing and fact checking; to L. Lieberman for his wise counsel; and to C. Ferland, M. Schuman, K. Bray, and C. Guillette.

A number of people helped with language translations: W. Smith and L. Hill (Chinese); A. McCormick and C. Stancioff (French); T. Hayes (Latin); Anna Booth and Erica Walch (Italian); and K. Hardy (Wabanaki). Many of my questions on the haiku of Issa and Buson were thoughtfully answered by D. G. Lanoue and J. Reichhold.

A heartfelt thanks to those who have accompanied me throughout my journey through illness or joined me along the way for the side trips. Please know how appreciated you are and that this book would not have happened without you. Some of you have a rare ability to understand and accept the invisible, and I could not have survived without that support: S. Tullberg, D. Lamparter, S. Spinney, L. Maria, A. Swan, and two truly exceptional physicians, Dr. C. Rosen and Dr. D. Bell.

Lastly, to all the creatures who at one point or another have shared their lives with me, including the snail and its 118 offspring, my deep *Homo sapiens* thanks.

Appendix: Terraria

I always keep several terraria of woodland plants in the house year-round. If you want to make a plant terrarium you can use any glass container or jar. Keep in mind that mosses, ferns, lichens, and other woodland plant species are often slow growing. Taking a few plants from a large patch of an unprotected species, on land you own or have permission to dig on, is usually fine. However, if you don't know your plants well, please consult a botanist first to make sure you are not taking species that might be endangered and protected by law. Alternatively, woodland plant material can be acquired from horticultural suppliers who specialize in their propagation. If the terrarium is to be inhabited by a living creature, make sure the propagated plants are organically grown.

Loam from the woods usually contains eggs of some creature or another, so occasionally an unexpected new friend may hatch out and surprise you.

Much can be learned from observing a snail where you find it, and letting it continue on with its life, undisturbed. If you choose to keep a snail for a little while, please give it the most natural home possible in a quiet location and provide fresh water and its familiar diet. Treat your snail gently, minimize handling, and return it to the same location where it was found, within the same season. I was glad when my snails were returned to their native habitat, and while I loved having them stay with me, I am most comfortable knowing that wild creatures are in their wild environments.

Selected Sources

Books on Gastropods

Barker, G. M. *The Biology of Terrestrial Molluscs*. New York: CABI, 2001.

———. *Natural Enemies of Terrestrial Molluscs*. New York: CABI, 2004.

Burch, John B. *How to Know the Eastern Land Snails*. Dubuque: Wm. C. Brown, 1962.

Chase, Ronald. *Behavior and Its Neural Control in Gastropod Molluscs*. New York: Oxford University Press, 2002.

Goldsmith, Oliver. "Of Turbinated Shell-Fish, or The Snail Kind." In *A History of the Earth and Animated Nature*. 1774. Glasgow: Blackie and Son, 1860.

Poe, Edgar Allan. *The Conchologist's First Book*. Philadelphia: Haswell, Barrington and Haswell, 1839.

Solem, Alan. *The Shell Makers: Introducing Mollusks*. New York: John Wiley and Sons, 1974.

Sturm, C. F., T. A. Pearce, and A. Valdés. *The Mollusks: A Guide to Their Study, Collection, and Preservation.* Boca Raton: American Malacological Society / Universal Publishers, 2006.

Wilbur, Karl M., ed. *The Mollusca.* 12 vols. New York: Academic Press / Harcourt Brace Jovanovich, 1983–88.

Wood, Searles, quoted in John Gwyn Jeffreys, *British Conchology, or An Account of the Mollusca Which Now Inhabit the British Isles and the Surrounding Seas.* Vol. 5, *Marine Shells and Naked Mollusca to the End of the Gastropoda, the Pteropoda, and Cephalopoda.* London: John Van Voorst. 1862.

Articles on Gastropods

Angelita, Giovanni Francesco. "On the Snail and That It Should Be the Example for Human Life." In *I pomi d'oro.* 1607. The Getty Research Institute, Research Library, Special Collection and Visual Resources, Los Angeles, CA.

Brieva, A., N. Philips, R. Tejedor, A. Guerrero, J. P. Pivel, J. L. Alonso-Lebrero, and S. Gonzalez. "Molecular Basis for the Regenerative Properties of a Secretion of the

Mollusk *Cryptomphalus aspersa*." *Skin Pharmacology and Physiology* 21 (2008): 15–22.
Chase, Ronald. "Lessons from Snail Tentacles." *Chemical Senses* 11, no. 4 (1986): 411–26.
———. "The Olfactory Sensitivity of Snails, *Achatina fulica*." *Journal of Comparative Physiology* 148 (1982): 225–35.
Cowie, Robert H., and Brenden S. Holland. "Dispersal Is Fundamental to Biogeography and the Evolution of Biodiversity on Oceanic Islands." *Journal of Biogeography* 33 (2006).
Dundee, D. S., P. H. Phillips, and J. D. Newsom. "Snails on Migratory Birds." *Nautilis* 80, no. 3 (January 1967): 89–92.
Gittenberger, E., D. S. J. Groenenberg, B. Kokshoorn, and R. C. Preece. "Biogeography: Molecular Trails from Hitch-Hiking Snails." *Nature: International Weekly Journal of Science* (January 25, 2006). http://www.nature.com.
Head, Sir George, *Tour in Modern Rome*, quoted in "Snails and Their Houses."
Ingersoll, Ernest. "In a Snailery." In *Friends Worth Knowing: Glimpses of American Natural History*. New York: Harper and Brothers, 1881.
Johnson, George. "Art. II.—Shell Fish: Their Ways and Works," *Westminster Review* 57 (January 1852).

Knight, G. A. Frank. Lecture delivered to the Perthshire Society of Natural Science, quoted in "Reversed Shells in the Manchester Museum," by R. Standen. *The Journal of Conchology*, edited by William E. Hoyle, 1904–6.

Lemaire, M., and R. Chase. "Twitching and Quivering of the Tentacles during Snail Olfactory Orientation." *Journal of Comparative Physiology A: Neuroethology, Sensory, Neural, and Behavioral Physiology* 182, no. 1 (December 1997).

MIT News Office. "MIT's RoboSnails Model Novel Movements." September 4, 2003. http://web.mit.edu/newsoffice/2003/robosnail.html.

Nielsen, G. R. "Slugs and Snails." University of Vermont Extension, Entomology Leaflet 14. 1998.

Pearce, Timothy A. "Spool and Line Technique for Tracing Field Movements of Terrestrial Snails." *Walkerana*, 4, no. 12 (1990).

Rollo, C. David, and William G. Wellington. "Why Slugs Squabble." *Natural History*, November 1977.

Sandford, E. "Experiment to Test the Strength of Snails." Notes and Queries. *Zoologist: A Monthly Journal of Natural History* 10, no. 120, Third Series (December 1886).

Shaheen, N., K. Patel, P. Patel, M. Moore, and M. A. Harrington. "A Predatory Snail Distinguishes between Conspecific and Heterospecific Snails and Trails Based on Chemical Cues in Slime." *Journal of Animal Behavior* 70, no. 5 (February 2005).

Simonite, Tom. "Slime-Riding Strategy Developed for Intestinal Robot." NewScientist.com, September 2006.

"Snails and Their Houses." *All the Year Round* 43, November 10, 1888.

Miscellaneous Science Books and Articles and Other Interesting Sources

Barbero, F., J. A. Thomas, S. Bonelli, E. Balletto, and K. Schönrogge. "Queen Ants Make Distinctive Sounds That Are Mimicked by a Butterfly Social Parasite." *Science* 323 (2009): 782.

Cocroft, Rex. "Thornbug to Thornbug." *Natural History*, October 1999.

Darwin, Charles. "Molluscs." Chap. 9 in *The Descent of Man, and Selection in Relation to Sex*. 1871. Princeton, NJ: Princeton University Press, 1981.

———. "Absence of Terrestrial Mammals on Oceanic Islands." In *The Origin of Species by Means of Natural Selection*. 1859. New York: D. Appleton, 1900.

Darwin Correspondence Project Database. http://www.darwinproject.ac.uk. Letters: #1962, to P. H. Gosse; #1967, to W. D. Fox; #2018, to J. D. Hooker; and #3695, to C. Lyell.

Dawkins, Richard. *The Ancestor's Tale: A Pilgrimage to the Dawn of Evolution*. New York: Mariner Books / Houghton Mifflin, 2005.

DeBlieu, Jan. *Wind: How the Flow of Air Has Shaped Life, Myth and the Land*. Emeryville, CA: Shoemaker and Hoard, 2006.

Freedman, David H. "In the Realm of the Chemical." *Discover* 223, June 1993.

Gawande, Atul. "Hellhole." *The New Yorker*, March 30, 2009.

Heidmann, Thierry. "Darwin's Surprise." *The New Yorker*, December 3, 2007.

Huxley, T. H. *A Course of Elementary Instruction in Practical Biology*. London: Macmillan, 1902.

Keller, Helen. *The World I Live In*. New York: Century, 1908.

Kellert, Stephen R., and Edward O. Wilson, eds. *The Biophilia Hypothesis*. Washington DC: A Shearwater Book / Island Press, 1995.

Kirby, Rev. William. *On the History, Habits and Instincts of Animals*. The Bridgewater Treatises, Treatise VII. 1835. Philadelphia: Carey, Lea and Blanchard, 1837.

Moalem, Sharon. *Survival of the Sickest*. New York: William Morrow, 2007.
Nightingale, Florence. *Notes on Nursing: What It Is, and What It Is Not*. New York: D. Appleton, 1912.
Oken, Lorenz. *Elements of Physiophilosophy*. Translated by Alfred Tulk. Ray Society, 1847.
Shubin, Neil. *Your Inner Fish: A Journey into the 3.5-Billion-Year History of the Human Body*. New York: Pantheon Books, 2008.
Stauffer, R. C., ed. “Mental Powers and Instincts of Animals.” In *Charles Darwin's Natural Selection: Being the Second Part of His Big Species Book Written from 1856–1858*. Cambridge: Cambridge University Press, 1975.
Villarreal, Luis P. “Are Viruses Alive?” *Scientific American*, December 2004.
———. “Can Viruses Make Us Human?” *Proceedings of the American Philosophical Society* 148, no. 3 (September 2004).
———. “The Living and Dead Chemical Called a Virus.” 2005. http://cvr.bio.uci.edu/downloads/05_villa_livedead.pdf.
———. *Viruses and the Evolution of Life*. Washington, DC: ASM Press, 2004.
von Frisch, Karl. *A Biologist Remembers*. Translated by Lisbeth Gombrich. New York: Pergamon Press, 1967.

Weir, James. *The Dawn of Reason: Or, Mental Traits in the Lower Animals*. London: Macmillan, 1899.

Weisman, Alan. *The World Without Us*. New York: Thomas Dunne Books / St. Martin's Press, 2007.

Wilson, Edward O. *Biophilia*. Cambridge, MA: Harvard University Press, 1984.

Zimmer, Carl. "Part Human, Part Virus." *Discover*, September 15, 2005.

Permissions

The author is grateful to the following authors, translators, publishers, copyright holders, and others granting permission to use excerpts from the following works:

Edward O. Wilson. Reprinted by permission of the publisher from *Biophilia* by Edward O. Wilson, pages 11, 12, 22, 106, Cambridge, Mass.: Harvard University Press, Copyright © 1984 by the President and Fellows of Harvard College. All rights reserved. *(Pages vii, 17, 141, 159.)*

Rainer Maria Rilke. From *Letters to a Young Poet* by Rainer Maria Rilke, translated by Stephen Mitchell. Copyright © 1984 by Stephen Mitchell. Used by permission of Random House, Inc. *(Pages 1, 31, 163.)*

Kobayashi Issa. Translations by David G. Lanoue. From his Web site *Haiku of Kobayashi Issa*. http://haikuguy.com/issa/. *(Pages 3, 85, 95, 149, 155.)*

Kobayashi Issa. Haiku of Issa's ("the snail gets up / . . .") from *The Essential Haiku: Versions of Basho, Buson & Issa,* edited and with an introduction by Robert Hass. Introduction and selection copyright © 1994 by Robert Hass. Unless otherwise noted, all translations copyright © 1994 by Robert Hass. Reprinted by permission of HarperCollins Publishers. *(Page 9.)*

Elizabeth Bishop. Excerpts from "Giant Snail" from *The Complete Poems 1927–1979* by Elizabeth Bishop. Copyright © 1979, 1983 by Alice Helen Methfessel. Reprinted by permission of Farrar, Straus and Giroux, LLC. *(Pages 25, 64, 69.)*

A. A. Milne. "The Four Friends," from *When We Were Very Young* by A. A. Milne. Copyright © 1924 by E. P. Dutton, renewed 1952 by A. A. Milne. Used by permission of Dutton Children's Books, a division of Penguin Young Readers Group, a member of Penguin Group (USA) Inc., 345 Hudson Street, New York, NY 10014. All rights reserved. *(Page 26.)*

Emily Dickinson. Reprinted by permission of the publishers from *The Letters of Emily Dickinson,* Thomas H. Johnson, ed., Cambridge, Mass.: The Belknap Press of Harvard University Press. Copyright © 1958, 1986, The President and Fellows of Harvard College; 1914, 1924, 1932, 1942 by Martha Dickinson Bianchi; 1952 by Alfred Leete Hampson; 1960 by Mary L. Hampson. *(Page 37.)*

Billy Collins. "Evasive Maneuvers," from *Ballistics* by Billy Collins. Copyright © 2008 by Billy Collins. Used by permission of Random House, Inc. *(Page 48.)*

Patricia Highsmith. Excerpts from "The Snail Watcher" and "The Quest for *Blank Claveringi*" from *Eleven.* Copyright © 1945, 1962, 1964, 1965, 1967, 1968, 1969, 1970 by Patricia Highsmith. Used by permission of Grove/Atlantic, Inc. *(Pages 51, 120–1.)*

Hans Christian Andersen. From "The Snail and the Rosebush." *The Complete Fairy Tales and Stories* by Hans Christian Andersen. Published by Random House Children's Books, a division of Random House, Inc. *(Page 103.)*

Karl von Frisch. *A Biologist Remembers.* Translated by L. Gombrich. Oxford, 1967.

Originally published in German, *Erinnerungen eines Biologen,* 2nd ed., 1962, by Springer (Berlin). *(Page 111.)*

Yosa Buson. Translation by Janine Beichman. Reprinted by permission of Cheng & Tsui Company, Inc. from *Masaoka Shiki: His Life and Works* by Janine Beichman (Boston, Mass.: Cheng & Tsui Company, 2002). Copyright © 2002 Cheng & Tsui Company, Inc. *(Page 113.)*

Gerald M. Durrell. Excerpts from *Birds, Beasts and Relatives*. Copyright © 1969 by Gerald M. Durrell. Used by permission of Viking Penguin, a division of Penguin Group (USA) Inc. For Canada and UK, reproduced with permission of Curtis Brown Group Ltd., London on behalf of the Estate of Gerald Durrell. Copyright © Gerald Durrell 1969. *(Pages 123–24, 126–27.)*

Yosa Buson. Haiku translation from *The Path of Flowering Thorn* by Makoto Ueda. Copyright © 1998 by the Board of Trustees of the Leland Stanford Jr. University. All rights reserved. Used with the permission of Stanford University Press, www.sup.org. *(Page 129.)*

Koybayashi Issa. Translation from *Haiku*. Copyright © 1949, 1950, 1952 by R. H. Blyth. Reprinted by permission of The Hokuseido Press. *(Page 143.)*

And yet when love-making is not in question, the snail is by no means sociable, although [it has been] *. . . observed in one branch of the family, snails engaged in mutually polishing a neighbour's shell with the foot.*

—"Snails and Their Houses," 1888

Elisabeth Tova Bailey's essays and short stories have been published in the *Missouri Review*, *Northwest Review*, and the *Sycamore Review*. She has received several Pushcart Prize nominations, and the essay on which this book is based received a Notable Essay listing in *Best American Essays*. She lives in Maine. Her web site is

www.elisabethtovabailey.net

Center Point Publishing
600 Brooks Road • PO Box 1
Thorndike ME 04986-0001 USA

(207) 568-3717

US & Canada:
1 800 929-9108
www.centerpointlargeprint.com